Praise for *Profit*

"A highly readable overview of human economic history, from foraging to modern industrial society. It surpasses all others in richness of detail and attention to environmental consequences."

—Donald Worster, University of Kansas

"This book offers important messages—but also fascinating asides and illuminating statistics, as it tells what may be the central tale of the human story."

—Bill McKibben, author *The Flag, the Cross, and the Station Wagon: A Graying American Looks Back at his Suburban Boyhood and Wonders What the Hell Happened*

"Fascinating."

—*The Toronto Star*

". . . eye-opening . . . Sweeping in scope yet grounded in intriguing particulars, this offers fresh perspective on an economic system 'we cannot live with . . . and cannot live without.'"

—*Publishers Weekly*

"Excellent."

—Bart Hawkins Kreps, *Resilience*

"With knowledge, skill, and stories of inventors, entrepreneurs, and conservationists, [Stoll] traces developments in technology, transportation, energy, communication, trade, and finance."

—*Nature*

"A concise and interdisciplinary history of capitalism . . . an excellent read for history enthusiasts."

—*World History Encyclopedia*

"A story of incredible ingenuity and villainy that begins in the Doge's palace in medieval Venice and ends with Jeff Bezos aboard his own spacecraft. Mark Stoll's revolutionary account places environmental factors at the heart of capitalism's successes and reveals the long shadow of its terrible consequences."

—Ian Angus, *Climate and Capitalism*'s Ecosocialist Bookshelf

"Incisive and compelling … an enjoyable deep read."

—Ramya Swayamprakash, *H-Environment*

"[*Profit*] aptly addresses a series of issues related to environmentalism, yesterday and today . . . [Stoll] intelligently mixes the social, economic, and environmental facets altogether. Undergraduates and even non-academic readers would learn from it."

—Yves Laberge, *Electronic Green Journal* 1 (49)

"*Profit*'s focus is not on proposing legislation or social activism but, rather, on presenting stories about the past. Yet it draws motivation from the present and offers lessons for the future. It speaks to the power of economic motivations and the immediacy of economic interests. But it also speaks to the more diffuse power of mentalities, ideas, and cultural norms. That interplay is at the center of the connection between the environmental and the economic. *Profit* suggests a path for historians who seek to deploy their insights about the past to serve the present, connecting the sweep of ideas and the realities of economically motivated action."

—Ian Kumekawa, *Enterprise & Society* (2024): 1–9

"The idea of an environmental history of capitalism is a fruitful line of inquiry, and *Profit* will be the foundation of this literature. The book is commendable for its wholistic approach to a topic that has only been dealt with peripherally."

—Jason L. Newton, *Environmental History* 29(1) (January 2024): 220–222

"A well-organized overview of the themes in the history of capitalism. Above all, it offers us a view, both disturbing and welcome in equal measure, on how neoliberal capitalism has splintered social cohesion, including religious and environmental consciousness."

—Ulbe Bosma, *International Review of Social History* 68(3) (2023): 519–521

"Stoll's book offers the opportunity to better understand how this world came to be . . . [and] gives historical context for the ubiquity of plastics and disposable products, the rise of fossil fuels, the concept of planned obsolescence and corporations' powerful tools of propaganda."

—Kiley Bense, *Inside Climate News*

"A wide-ranging book that takes readers from antiquity all the way to today's internet-powered global consumerism, exploring the environmental consequences along the way . . . Interesting . . . Fascinating."

—Jeremy Williams, *The Earthbound Report*

"[A] revolutionary account [that] places environmental factors at the heart of capitalism's progress and reveals the long shadow of its terrible consequences."

—Michael Svoboda, *Yale Climate Connections*

Profit

Environmental History

Adrian Howkins, *The Polar Regions*
Paul R. Josephson, *Chicken*
Jon Mathieu, *The Alps*
Mark Stoll, *Profit*

PROFIT

An Environmental History

Mark Stoll

polity

First published in 2023 by Polity Press
This paperback edition published in 2024 by Polity Press

Polity Press
65 Bridge Street
Cambridge CB2 1UR, UK

Polity Press
111 River Street
Hoboken, NJ 07030, USA

ISBN-13: 978-1-5095-3323-7
ISBN-13: 978-1-5095-3324-4 (pb)

A catalogue record for this book is available from the British Library.

Library of Congress Control Number: 2022938554

Typeset in 11.5 on 14pt Adobe Garamond
by Cheshire Typesetting Ltd, Cuddington, Cheshire
Printed and bound in Great Britain by CPI Group (UK) Ltd, Croydon

For further information on Polity, visit our website:
www.politybooks.com

Contents

Illustrations

Acknowledgments

In writing this book, I have incurred many scholarly and personal debts. First I must mention my gratitude to my editor at Polity, Pascal Porcheron, who invited me to write it and who followed supportively through the entire process as the sausage was being made. Thanks also to the many staff at Polity with whom I worked in the production of the book.

I am grateful to Sverker Sörlin for reading and commenting quite helpfully on the first draft. John R. McNeill also commented with insight at various stages. I also appreciate the responses of several anonymous readers of the proposal and of the first draft.

A fellowship at the Rachel Carson Center of Munich, Germany, under the direction of Christof Mauch, in the fall of 2019 gave me time among supportive colleagues from around the world during which I wrote a portion of the book. Specifically, thanks to the fellows who at a work-in-progress session discussed and commented on an early version of a chapter: Liz DeLoughrey, Kelly Donati, Rachel S. Gross, Robert Gross, Marcus Hall, Sevgi Sirakova Mutlu, Eunice and Rubens Nodari, Anna Pilz, Jayne Regan, Xiaoping Sun, Anna Varga, Monica Vasile, and Kate Wright. The RCC fellowship program was a wonderful opportunity for scholarly creativity and community.

Heartfelt thanks to Verena Winiwarter for the invitation to present a version of the book's thesis at a Minisymposium at the Center for Environmental History (ZUG) in Vienna, Austria, in 2019. Thanks as well to the audience of the panel at the 2018 meeting of the American Society for Environmental History in Riverside, California, for their questions and comments on my paper on consumer capitalism and the environment. I also presented a version of the book's thesis to a session organized by Shen Hou and Donald Worster of the Center for Ecological History at the Renmin University of China, which formed part of the ASEH's Environmental History Week in 2021.

I owe a debt of gratitude to the History Department of Texas Tech University and its chair Sean Cunningham for release from teaching at important moments. The Texas Tech Humanities Center Faculty Fellows program funded a semester's course release. Thanks to several History Department colleagues, who cheerfully gave me advice and suggestions for readings when I ventured in fields far from my own: Stefano D'Amico, Abigail Swingen, and Barbara Hahn.

Finally, I benefited from correspondence with and advice from many colleagues. My appreciation goes to Karl Appuhn, Bruce Clarke, Craig Colton, Steve Epstein, David Fedman, Maria Margarita Fernandez Mier, Alan Loeb, Greg Maddox, Geneviève Massard-Guilbaud, Jim McCann, Manuel Gonzalez de Molina Navarro, Paul Warde, and John Wing.

Innocent Afuh and Cynthia Henry of the Texas Tech University Library helped me find answers to some difficult research questions. The library's reference staff often fielded requests for help finding sources. I kept Katie DeVet and the staff of the Interlibrary Loan office very busy. I am grateful to all of them for their cheerful and capable assistance.

While I am grateful to all these people for so much, the interpretations and any errors in the book are mine alone.

Most of all, I owe my wife, Lyn, a debt I cannot repay for her constant support and encouragement during the five years I wrestled with this monster of a book topic. This book is for her.

Introduction

Reach into your pocket or bag and pull out your smartphone. Almost certainly you have one within easy reach. You might even be reading this on it. The palm-sized technological marvel you hold can connect you to billions of people, even in some of the most remote spots on earth. It can tell you where you are and how to get to where you are going. It can play music, videos, and movies. With it, you can find almost any information you want, in many languages, from the history of the Wattasid dynasty of Morocco to Great-Aunt Tilley's latest pictures of her cat. The consumer products of the world's factories and artisans are yours with a touch of a "Buy" icon. This little device has wormed its way into daily life so deeply that separation from it provokes anxiety.

The smartphone is ubiquitous because it is cheap—so cheap that service providers give away older models. This miraculous contrivance, though, costs more than advertised. The touchscreen and case conceal a Pandora's box of environmental evils. The plastic in the case derives from petroleum or natural gas, which a multinational corporation extracted from underground, often with damage to ecosystems and watersheds, and transported by pipeline, supertanker, truck, or railroad with spills, leaks, and clouds of the greenhouse gases methane and carbon dioxide. A low-paid worker in a factory in east or south Asia used dangerous chemicals to make and form the plastic case. Inside the case, metals like rare earths, aluminum, gold, cobalt, tin, and lithium make the phone work. The rare earths come from China, which has 95 percent of the world's reserves. Mining them uses tremendous amounts of energy and materials and generates radioactive waste, hydrogen fluoride, and acidic wastewater. Ships burning fossil fuels bring the other metals from mines in southeast Asia, Africa, South America, and Pacific islands. Those mines have displaced people, incited violent conflict, destroyed agricultural land, polluted water, and damaged the health of often unprotected miners, who sometimes are child laborers. Energy for manufacturing your phone and charging its batteries (not to mention powering telephone and Internet services and server farms) comes, more often than not, from powerplants that burn coal or gas and produce sulfur and mercury pollution and more greenhouse gases. Companies design smartphones to be replaced about every two years. Most people discard old phones, which end up in landfills, slowly leaching toxic chemicals. Some owners recycle them, but recycling their complex, compact, and integrated components is not easy or clean and requires harmful chemicals that produce hazardous waste.[1]

How did the environmental crime that is the smartphone end up in the hands of billions of people who have little idea of its environmental costs? The answer is complicated. Most certainly, people want them for convenient communication, social media, entertainment, features like cameras and alarms, and access to the Internet. They may be responding to social pressure or work requirements. Then, too, smartphones have nearly become a necessity in modern life. But looming behind all these factors is a group of huge corporations that sell them for the profit of investors and shareholders. Other corporations provide the content, application programs ("apps"), and social media websites that cleverly keep users' attention nearly constantly on their phones. Finally, standing almost hidden in their shadows are some of the most important actors, the corporations that lobby governments for favorable laws and regulations and others that provide promotional campaigns to convince people to buy the device or its attendant services and apps. They have been extraordinarily successful at getting people to acquire something that only a short time ago no one knew they wanted. Before 2007, the smartphone barely existed, but in the near future, 70 percent of the world's population will have one. The smartphone is the very epitome of capitalism in its latest incarnation.

Who then is guilty of the environmental crime that is the smartphone? A detective on the case might suggest the *Murder-on-the-Orient-Express* solution: everybody did it. Under the promptings of a variety of considerations, we thoughtlessly buy and dispose of smartphones without any compulsion whatsoever. Alternatively, an environmental Hercule Poirot might instead name the culprits who most stand to benefit and who hide the smartphone's environmental costs: the corporations that profit from making and selling the device and its services. The smartphone and its cousins the tablet, laptop, and personal computer are astonishingly profitable. They provide the incomes of what in April 2021 were the world's six largest private corporations by market value: Apple, Microsoft, Amazon, Alphabet (Google), Facebook, and Tencent. Founded since 1975, and four of them since 1994, these gigantic corporations are barely older than the smartphone itself and none is older than the personal computer.[2]

We have lined up suspects for the crime. What will be the verdict? We have a hung jury. Some jurors call the crime a part of the "Anthropocene" and blame humans collectively. Environmental thinkers like Bill McKibben, E. O. Wilson, and Elizabeth Kolbert blame our personal choices.[3] If everyone would simply behave, believe, buy, and ballot more responsibly and more morally, we could save the world. Others, among them Jason W. Moore, Andreas Malm, and Naomi Klein, prefer "Capitalocene" and point the finger of guilt at capitalists and corporations.[4] Jurors also disagree over the time of the crime. One party dates it to a Great Acceleration after World War II. Another argues for the beginning of the Industrial Revolution around 1800. Still a third group sees human fingerprints all over the environment much deeper in time, perhaps as long ago as the adoption of agriculture 10,000 years ago.

Frankly, the fault lies both in our stars and with ourselves. It is difficult to identify anything in our modern lives—food, fuel, clothing, housing, transport, work, or leisure—that is not also complicit in similar crimes against the earth. They all present our jurors with the same conflicting evidence. One may argue that the rich, who consume the most, are more culpable than the poor, and that no amount of water can wash guilt from the hands of leaders of industry. True enough. Yet none of us today lives completely free of the web of consumer capitalism. The ongoing pillage of the planet lends length, ease, and quality to the lives of the great majority of us and in truth makes it possible for us to exist at all. The human species flourishes as no biped or quadruped ever has. Neither hunting and gathering nor premodern agrarian practices, as sustainable as they often are, could feed, clothe, and house the world's population of about 8 billion people or give a substantial minority of them previously unimagined abundance, but the global capitalist system does.

Calculations of the total body weight of all humans and land animals through human history show how miraculous this really is. In the late Pleistocene, according to one study, human biomass was vanishingly small, while terrestrial mammals collectively weighed about 175 million metric tons.[5] By 1900, after the agricultural and industrial revolutions, as Vaclav Smil has reckoned it, humans weighed in at 70 million metric tons while wild animals had declined to 50 million metric tons. Species under human control living on formerly wild habitat—cattle, horses and donkeys, water buffalo, pigs, sheep, goats, camels, poultry, and more—counted for 175 million metric tons, the total body weight of all wild animals many millennia earlier. In other words, humans had found ways to support an additional 120 million metric tons of body weight out of the same environmental resources.

Then consumer capitalism kicked in and terrestrial biomass made an astounding jump. By 2000, Smil calculates, humans alone made up over 300 million metric tons of biomass, or over two-thirds more animal weight than all animals together before the rise of *Homo sapiens.* Domesticated animals composed another 600 million metric tons (with a couple million more for dogs, cats, and other pets). The combined biomass of humans and domesticated animals amounts to greater than 500 percent of the Pleistocene total. At the same time, the biomass of wild animals had declined another 50 percent since 1900 to about 25 million metric tons, about a sevenfold decline since their Pleistocene peak. This means that the figure for wild animals equates to about four percent of that for domesticated animals and only 2.7 percent of that for humans and their animals together.[6]

Perhaps we should be grateful that we have somehow left any resources for them at all. But this marshalling of the earth's resources for human needs is what makes it possible for about 8 billion people to live.

This book tells the story of how this world came to be. The title *Profit* is ambiguous. In its common current usage, it means financial profit, which today we associate with a system we call capitalism. Profit in its original meaning can also mean any benefit, as when we say we profit by experience. In the financial sense, most of capitalism's profits accrue to a very small proportion of the world's population. In the broader sense, considering that capitalism supports so many people, and a large proportion of us very comfortably, we all profit from it. *Profit* tells the history of capitalism through the stories of a series of individuals who represent either the opening of a significant new stage of capitalism or the development of influential movements to control its environmental impacts. Many (but not all) of the former believed they were profiting society and themselves at the same time. The latter sought to profit the world without profits to themselves.

Capitalism has transformed many times throughout history. At its most fundamental level, capitalism is an economic system in which (usually) private owners of accumulated wealth or capital invest it for profit in enterprises of extraction and distribution of raw materials or of production and distribution of goods, where (ideally) unrestricted competition with other enterprises in a (more or less) regulated marketplace decides prices. Note the qualifiers in parentheses. They suggest how rare anything like "pure" capitalism has been and help account for significant variations from place to place and era to era. A capitalist system functions more effectively when it has access to easy transportation, abundant energy sources and natural resources, a tractable and disciplined labor force, and dependable communication between extractors, producers, and sellers.[7] Its scale runs from the mundane, such as when a tenant farmer sells papayas in a market in a tropical village, to the momentous, as when a wealthy trader sells arcane financial instruments in a global center of capital like New York or London. Between local and global levels are many almost incomprehensibly complex and interconnected layers, always in motion and changing in response to the others. The price farmers get for commodities like papayas affects global capitalism, while tropical farmers feel the impact of international financial machinations.[8]

Capitalism's development was anything but preordained. Constant motion and change in the layers and interplay between them have driven capitalism's historical evolution from incipient to modern stages. Its history is contingent, culturally conditioned, and in constant conversation with environmental possibilities and constraints. Recognizably capitalist systems evolved not just in western Europe but around the world as well. Only in the West, however, did modern capitalism arise, with its hallmarks of separation of business from the household, rational bookkeeping, and rational organization of labor. Neither of two major historical forms of capitalist organization, the slave plantation and the mechanized factory, had any true counterparts outside modern capitalism's realm. Western capitalism wields unprecedented power to utilize resources intensively to produce wealth. Because industrial and consumer capitalism developed in western Europe and North America and today reaches into every corner of the globe, this history focuses mainly on the paths that led to, through, and out of these regions.[9]

Capitalism has no single place or time of origin. Every form of capitalism had forerunners centuries earlier. Contemporary consumer capitalism developed mainly in the United States from precursors in the European and American Industrial Revolution. Most elements of English industrial capitalism developed from innovations in the Netherlands a century before. For their part, the Dutch had poached Iberian global empires but ultimately relied on a model of mercantile capitalism that developed in medieval Italian trading empires. Italians had in turn borrowed prototypes devised in the Islamic world and ancient Rome and Greece. From Antiquity the trail leads back into the deep past. The first simple strands of the worldwide web of consumer capitalism that ensnares us today were laid down many millennia ago and can be discerned even among our earliest ancestors. Its threads would be spun, broken, repaired, abandoned, replaced, extended, and multiplied many times, until today they entangle all the earth and its peoples.

Some threads remained constant from the Age of Sharp Stones to the Age of Smartphones. From the time humans first appeared on the planet, they showed a talent for extracting more sustenance from a region's resources than any other animal and could support greater numbers, greater biomass. They communicated and cooperated better than any other animal. Humans mined minerals and soils for utility and for consumer goods. They sought sources of more energy. They manufactured useful and desirable items. They devised more efficient transportation for goods and for themselves. Humans dominated, exploited, and fought with their neighbors. And they developed the earliest economy with their exchanges of goods, resources, skills, and knowledge.

Capitalism's story is tightly woven together with the natural world. On the one hand, specific environmental circumstances made modern capitalism possible and shaped its growth. Broader climate changes also played a significant role. On the other hand, economic activity has always degraded environments. With each stage of economic development, people ratcheted up the efficiency of their exploitation of natural resources. Step by step, the process impoverished ecosystems and transformed landscapes. Today, consumer capitalism has magnified human environmental impact to a startling degree, exterminating species, deranging ecosystems, draining wetlands, redirecting and damming rivers, cutting down forests, and degrading soils. It consumes resources and dumps them again with dizzying speed. It broadcasts chemicals to the four winds, the four corners of the globe, and the seven seas. It changes the atmosphere and heats the planet. There's hardly a place or species on earth unaffected by human deeds and works—even to the bottom of the deepest ocean trenches, where high levels of toxic chemicals poison creatures and plastic bags float like pale ghosts.[10]

Profit: An Environmental History traces how these developments in technique, technology, transportation, energy, communication, and trade and finance led to modern consumer capitalism. It presents incipient capitalism, mercantile capitalism, plantation capitalism, industrial capitalism, and consumer capitalism as stages in the long human endeavor to use resources more intensively. Each chapter explores a stage in its evolution and its environmental impacts. Chapter 1, "How It Started," gives an account of precursors and emerging patterns among early hominins. It then explores ancient Greek and Roman merchant capitalism, with its fateful linkage of coinage, slavery, commodity agriculture, trade, empire, and environmental change.

Chapter 2, "Trade and Empire," tells of the signal advances by medieval Italian city-states in banking, corporate institutions, empire, and commodity agriculture. Italians fashioned the template that Iberians and northwest Europeans followed with their own world trade and plantation empires. The chapter uses the experience of Christopher Columbus to illustrate the transfer of Italian models, as well as the haphazard, opportunistic methods that built European imperial systems and economies. Global exploitation of peoples and nature fueled Europe's rise.

The next two chapters, "The Wonders of Coal and Machines" and "Age of Steam and Steel," trace the adolescence and maturity of industrial capitalism, often called the First and Second Industrial Revolutions. This era ushered in tremendous, fateful technological, social, and environmental changes, among them the energy transition from renewable to fossil energy. The careers of James Watt and Andrew Carnegie illustrate important themes, among them the role of science in technological advances, cultural influences on the pace and direction of economic and social change, and the passing of industrial leadership from Europe to the United States. Chapter 5, "Conserving Resources," considers the emergence of a conservation and parks movement to control industrialization's environmental problems. American George Perkins Marsh and Englishman William Stanley Jevons perceptively and influentially analyzed industrialization's destruction of its resources and energy bases.

Three chapters, "Buy Now—Pay Later," "Stepping on the Gas," and "Selling Everything," portray the evolution, mainly in the United States, of twentieth- and twenty-first-century consumer capitalism. This form of capitalism bears more responsibility than anything else for today's environmental crises. Advances in industrial techniques had pushed production past demand. To boost consumption, corporate management developed new techniques in advertising, financial incentives, and product replacement. Another energy transition, from coal to oil and gas, fueled both production and consumption. Supply and demand chased each other at a rising pace. Our houses filled with stuff. Global wealth rose, too. Nothing exemplifies the age like the automobile industry, as the new consumerist orientation of Alfred Sloan's General Motors outstripped the older producerist mentality of Henry Ford's Ford Motor Company. After World War II, Ray Kroc represented a new type of capitalist figure, cut from a different cultural cloth with fewer moral qualms about having money. Kroc masterminded the global rise of the model fast-food restaurant chain, McDonald's. Finally, Jeff Bezos's Amazon.com exemplifies the period of consumer capitalism after the 1970s, in which selling things to consumers has become much more profitable than making them. Rising levels of consumption draw producers to make ever more goods, for which they take ever more resources from the environment and generate ever greater quantities of waste and greenhouse gases.

"The Rise and Globalization of Environmentalism" explores the global movements to manage consumer capitalism's terrible ecological impact. American Rachel Carson's *Silent Spring* stands as the single most significant work in the postwar environmental movement. Briton Barbara Ward's *Only One Planet* (cowritten with René Dubos) and her tireless work around the world laid out major themes in the globalization of environmentalism. The two books illustrate environmentalism's major concerns but also its important limitations. Environmentalism remains entangled with consumer capitalism even as it works to counter its ecological impacts. A conclusion offers reflections on how our efficiency in extracting resources from the natural world has far outrun our ability (or willingness) to manage environmental consequences.

Although capitalism has long interested environmental historians, they had never made it the subject of a monograph or the primary actor or theme of a major work. Such foundational books as Donald Worster's *Dust Bowl* (1979) and William Cronon's *Changes in the Land* (1983) stressed such abstract forces as Worster's "culture of capitalism" or Cronon's "commodification" but did not analyze the system itself. The economic crisis of 2008 and the appearance of much-discussed books like Thomas Piketty's *Capital in the Twenty-First Century* (2013) and Sven Beckert's *Empire of Cotton: A Global History* (2014) prompted renewed discussion of capitalism and its impacts. A stream of works has begun to address the subject, prominent among them the works of Naomi Klein, Jason Moore, and Andreas Malm.

Profit: An Environmental History takes a different tack. It is a history of capitalism that seeks to explain both how capitalism changed the natural world and how the environment shaped capitalism. Capitalism grew up with our species and is inescapably everywhere. It has sinuously and stealthily worked its way into and subverted every system human ingenuity has thus far devised to control or destroy it. *Profit* tells the development of the conundrum that capitalism presents us today: we cannot live with capitalism and cannot live without it. At best, we can work to ameliorate its worst effects.

A book of this length must leave out topics that I would love to address more fully. Perhaps if I had the luxury of writing three volumes, as Fernand Braudel did (although even he kept mainly to capitalism in a few European countries in just the fifteenth through eighteenth centuries), I could devote greater attention, for example, to capitalist systems in China, India, and other civilizations. Also, this book discusses Western environmental thought but gives little explicit attention to economic thought, whether of Adam Smith, Karl Marx, Joseph Schumpeter, Karl Polanyi, Milton Friedman, or any other. Debates among economists over the environment grew so voluminous in the last century that they would fill a book in themselves. Finally, I allude to the experience of worker, subaltern, and indigenous groups but do not focus on them, which again are topics worthy of another book. That still leaves me with plenty to write about.

ONE

How It Started

An ecological odyssey

As the 1968 movie *2001: A Space Odyssey* opens, humanity's ape-like ancestors three or four million years ago are struggling to survive. One day, unseen aliens place a large black monolith in their midst. Inspired by its mysterious powers, the hominins learn to use tools. They first wield weapons against prey and predators, ensuring their survival. Soon, however, they murderously turn them on their neighbors, driving them away from territory and resources. The monolith vanishes, but a second one lies buried on the moon, whose purpose is to send a beacon when humans have advanced enough to find it and expose it to the light of the sun. The surviving crewmember of a spaceship sent to follow the direction of the beacon discovers a huge third monolith orbiting Jupiter. This one is a Star Gate, which takes him across the galaxy to a distant planet. There, a fourth monolith transforms him to the Star Child. His evolution complete, he returns to earth to bring peace to warring nations.[1]

However much humanity could use a little benevolent extraterrestrial guidance these days, we unfortunately have found no alien monolith buried on the moon for uncounted eons to guide us out of a troubled present into a utopian future. On the other hand, in the decades since author Arthur C. Clarke and director Stanley Kubrick wrote the novel and screenplay, we have come to realize there was no need for aliens to have intervened in human history to set humanity on its present track. Observers have watched many ape, monkey, and bird species use tools. Archeologists discovered evidence that pre-*Homo* hominids also used stone tools and seem to have figured out tool-making on their own. (On the other hand, a utopian future still seems likely to require outside assistance.) Our ancient ancestors also, stone by stone, laid the foundation for the rise of early capitalism, changing the environment as they did so.

In the beginning: economy and ecology

When the earliest species of the genus *Homo* evolved about 2.6 million years ago, something happened almost as dramatic as a timely hint from benevolent aliens. Developments having no equal in earthly evolution set our ancestors on the long, meandering course that has led us to today's global consumer capitalism and global environmental crisis. The first steps were small but fateful. Superior powers of communication made possible greater cooperation. Our omnivorous ancestors now gathered and used resources for their survival more effectively. They passed accumulated cultural and technical knowledge to others and to future generations. Greater manual dexterity allowed them to improve the basic spears and stone tools of chimpanzees and *Australopithecus.* Talking and working together, hominins could now bring to the table animals as large or larger than they. Defending their kills against scavengers and thieves, they took their place among top predators. A meatier diet with considerably more protein made them larger and perhaps more intelligent.

Early humans also fatefully made the first energy transition and learned to get more out of available natural resources. They tamed a vital source of energy to supplement muscle power when they learned to liberate the heat energy stored in plants. Humans mastered fire possibly 1.9 million years ago, probably by 790,000 years ago, and certainly by 300,000–200,000 years ago. Fires burned undergrowth and made forests more open. Game species thrived, and with them, humans. For the first time in the history of life, a species deliberately refashioned whole ecological systems to support more of its kind from the same physical resources. Cooking food gave hominins access to a wider selection of food sources and, it has been suggested, allowed energy saved in digesting raw foods to feed the voracious energy appetite of larger brains.[2] Fire kept humans warm as climates cooled after they spread into temperate latitudes during warm interglacial periods.[3] Humans gave vent to creative impulses with fired clay figures at some early date, followed by pottery and, about 10,000 years ago in the early Holocene, by copper, all gifts of the stored energy of burning wood.

While the durability of stone implements lent the name "Stone Age" to the era, early humans developed all sorts of technology and simple machines. Bones made excellent material for such tools as awls and needles. No doubt early humans used perishable body parts like sinews as well. About 70,000 years ago, cooling climates in once-warm latitudes into which hominins wandered inspired leather-working technology for clothes, bags, and other objects.[4] At some time before the onset of the Holocene about 11,700 years ago, humans developed bows and arrows, fishhooks, and throwing-sticks for hurling darts, spears, and harpoons. Seaworthy boats, probably built for fishing the ocean, carried people across fifty miles of open water to Australia 65,000 years ago.[5] People wove dyed fibers into textiles for stylish colorful clothes no later than 30,000 years ago.[6] Basket-weaving provided storage until fired clay pots offered all-purpose waterproof and vermin-proof containers as early as 20,000 years ago.[7]

So it was that the earliest human species developed technology and simple machines and harnessed stored energy—the tiny mustard seeds of the Industrial Revolution. Technology and energy are, however, not sufficient in themselves. The extraordinarily long infancy and childhood of humans gives time for teaching and training at a unique scale. Humans organized themselves to kill large prey, cook, and share food, perform ceremonies, and much more. They developed techniques to produce dangerous weapons, hunt, form pots, and kindle fires. Far beyond that, they also had techniques for binding wounds, treating illness, and using magic and ritual to attempt to control for human benefit the mysterious forces that rule the world.[8]

The environmentally destructive mining that brings us the components of our cell phones is an ancient practice writ large. Humans are a mining species. The nearest rock was likely not the best rock for human purposes. Early humans dug shallow open-pit mines to get to unweathered flint, which flaked better. Mining always leaves scars upon the land, a boon for archeologists. Evidence survives of a 1.3-million-year-old flint quarry in Morocco.[9] In Swaziland at least 43,000 and perhaps 80,000 years ago, miners extracted tons of specularite (a source of red ocher). By 35,000 years ago, people in Poland and Hungary had dug pits and sometimes subterranean mines for colored ores, while in Paleolithic Egypt, Australia, France, Spain, Belgium, Poland, and (later) in Texas, they dug up flint.[10]

Groups that controlled mines found themselves in economically valuable positions, direct ancestor of today's mining, manufacturing, and trading enterprises. Archeologists have found obsidian and red ocher far from the nearest outcrop and inferred development of kinship, social, and trade networks.[11] Groups mined and manufactured tools and exchanged them for articles they did not have to hunt or make themselves, presumably at an advantage to themselves (the first profits). The market for red ocher represents the earliest known demand for consumer goods, which had no practical use. Scattered surviving evidence of beadwork, pigment from red ocher, and other artifacts suggests the development by 300,000 years ago of status markers, symbolic thought, and perhaps religion. By late prehistoric times, desirably distinctive stone spread along trade routes hundreds of miles from the mines it came from.[12]

When *Homo sapiens* first appeared, between 200,000 and 300,000 years ago, they prospered, went forth, and multiplied. Population grew very rapidly—it leaped tenfold in just the transition from *H. neanderthalensis* to *H. sapiens* in Europe around 40,000 years ago. In multiplying so rapidly, they repeatedly overdrew the fund of resources landscapes offered to foragers. Some of them would have to move elsewhere, pulled by greener pastures and pushed by stronger competitors. Between 120,000 and 90,000 years ago, wet and cool conditions enticed *sapiens* into the Levant and Arabia. When the climate dried 65,000 years ago, *Homo sapiens* rolled across the world in a global blitzkrieg.[13] They soon reached Australia. No later than 15,000 to 12,000 years ago humans stepped on American soil. No other vertebrate has managed to thrive in virtually every ecosystem from tropics to tundra (and now ventures up into space and down to the bottom of the sea).[14]

Whenever we might date the onset of the Anthropocene, *anthropos* changed environments and left a mark on the fossil record wherever they went. Escaped cooking fires and deliberately set fires announced the ability of this new-and-improved ape to alter ecosystems, sometimes radically.[15] No later than two million years ago, competition from these new top predators caused the rich diversity of large African predator species to decline for the first time.[16] Hominins also suppressed the diversity of African species of large mammals.[17] When *Homo sapiens* charged across the world's continents, they simplified (or, really, impoverished) ecosystems as they went. The Pleistocene had fostered an unusually abundant and diverse array of extremely large mammals on every ice-free continent: mammoths, giant ground sloths, saber-toothed cats, hippopotamus-sized wombats (*Diprotodon*), and many more. Climate change exacerbated the stress animals were under, of course, but never had climate pressure alone selectively eliminated only large animals. But now, as humans arrived conversing, cooperating, making weapons, setting fires, and reproducing abundantly, megafauna went extinct everywhere.[18]

Hominins simplified ecosystems by removing species but also rearranged ecosystems by introducing alien species, a propensity that today has altered nearly every ecosystem on earth. Evidence in southeast Asia shows that prehistoric foragers translocated plants and animals far from their ordinary habitats. No later than 45,000 years ago, yams appeared on both sides of Wallace's Line dividing the flora and fauna of Asia and Australia. Humans translocated animals and plants as they migrated and surely also traded species as they exchanged information and goods with others. They seem also to have encouraged the growth of food trees near their settlements, possibly by planting nuts and seeds. Transplantation and cultivation long pre-dated the Agricultural Revolution in the Near East.[19]

At the same time, like the primitive hominins in *2001*, humans also turned their aggressive capacity for communication and cooperation against competitors of their own genus. Practices that evolved into war and slavery began at a very early date. Several species of hominins had always coexisted at any time. Now, however, when *Homo sapiens* moved into the neighborhood, neither man nor beast was safe. Neanderthals, Denisovans, Floresians, and all other species of *Homo* died out when cousin *H. sapiens* arrived. The inclination for violence strengthened through time as numbers increased and weapons grew deadlier.[20] The slaughter of people on a huge variety of pretexts suggests the power of human tribalism and hostility to difference. Evidence from more recent times repeatedly shows that groups that controlled important resources attracted conflict with envious others, particularly during climate stress and other crises.

People very early in human history must have faced the question of what to do with captives. If the practices of pre-Columbian Americans are any guide, societies treated captives variously: adoption into the tribe, servitude including slavery, or death by torture or sacrifice. The recent discovery that small amounts of DNA from other species of *Homo* survive in our bodies suggests that bands of *Homo sapiens* likewise captured vanquished foes and incorporated some of them into their groups, by either adoption or enslavement.[21]

By the end of the Ice Ages, a once meek denizen of field and forest had inherited the earth. *Homo* altered ecosystems and reduced diversity and body size of other species. By the time humans had crossed the oceans to Madagascar and the islands of the Pacific, transforming ecosystems and extinguishing species as they went, humanity, who were once merely bad neighbors, had become bad landlords. Just as importantly, humans had carried fundamental elements of modern capitalism—technology, machines, use of concentrated energy, efficient exploitation of resources, mining and manufacturing for use and consumption, trade, competition, conflict, domination, ecological disruption—to the ends of the inhabitable earth.

Planting and herding: inventing capital and property

Humans began to practice true pastoralism and agriculture with the onset of the Holocene. They now passed irrevocably through a gate that opened onto momentous changes in society, economics, and the planet. Agriculture created surpluses that led to towns and cities. Literacy and literature developed. Power and wealth accumulated. Trade increased. Towns and then empires warred with each other. Inequality and slavery grew dramatically. Both herding and farming radically simplified ecosystems for human benefit. Moreover, they altered the climate.

Why did *Homo sapiens* around the world wait until the Holocene to develop agriculture independently in many places? The earliest agriculture known to us were the yam fields of the tropical Borneo highlands 30,000 years ago. Grinding stones and hearths from Kebaran settlements 19,000 years ago on the shores of Lake Galilee, and from the bread-baking Natufian culture throughout the Levant between 14,000 and 11,000 years ago, attest to preparation of nuts and wild cereals that could not be consumed unprocessed.[22] These were notable exceptions. Perhaps climate instability during the late Pleistocene discouraged successful agriculture elsewhere. Conditions swung back and forth from warm to cold and wet to dry, often very suddenly. The onset of the Holocene inaugurated an unusually stable climate that apparently encouraged agriculture.

Agricultural life did not charm foragers out of the woods—resource dearth pushed them. Eating bread in the sweat of one's face, as Genesis puts it, had little appeal if land provided sufficient food for the relatively low investment of effort of hunting and gathering. Clearing, planting, harvesting, and grinding is hard work.[23] In the more stable Holocene climate, human population grew more quickly. Extinction of large mammals took a lot of protein off the menu. Hunted more intensively, surviving game grew scarcer. Herding, horticulture, and agriculture provided more food for more people from the same resource base by reducing the biodiversity of an ecosystem to a smaller number of species of edible or useful animals or plants.

If space was available, herding ensured a reliable food supply with less labor than full-time farming. Alfred Crosby has pointed out how herds turn grass, stalks, and chaff, which humans could not eat or use, into steaks, chops, milk, and leather.[24] Early in the Holocene, humans domesticated cattle in upper Mesopotamia. Later, Egyptians domesticated donkeys and Indians zebu cattle.[25] Milk and milk products like yogurt or cheese provided protein without destroying the animal. Pastoralism, though, transformed social and power relations. Herds allowed people to accumulate wealth. (The English words *capital*, *chattel*, and *cattle* share a Latin root.) Groups moved toward privatization of resources. Like the biblical patriarchs, herders reckoned their wealth in livestock, slaves, and offspring. Job, for example, counted "7000 sheep, and 3000 camels, and 500 yoke of oxen, 500 she asses, and a very great household," with seven sons and three daughters (Job 1: 2–3). Where herders went, greed, envy, theft, and violence followed. Private property is made to be stolen. Tales of cattle theft abound across pastoral cultures, such as Indra's rescue of cattle from the Panis, Herakles' labor of stealing the cattle of Geryon, Queen Medb's raid to capture the bull of Cualnge, or David's sneak attack to slaughter Philistines and plunder livestock (1 Samuel 27: 9).

People without extensive grasslands or in dense populations adopted farming instead. The human relationship with nature changed. Nature worked constantly against artificially simplified ecosystems. Where people hoed, bare and broken ground brought forth thorns and thistles and in sorrow the sons and daughters of Adam ate from it. Fields planted with a single species presented a banquet to insects, birds, rodents, and plant diseases. Farmers battled constantly to subdue nature and have dominion over it. The egalitarian hunter-gatherer-universe of humans, animals, and spirits departed. Capricious sky and nature gods filled farmers' cosmos. Rituals for fertility (both sexual and agricultural) proliferated.

Early farmers diversified to expand and enrich their diets and hedge against epizootics and bad harvests. They kept herds. During the Chalcolithic Age, which began about 8,500 years ago, Near Eastern farmers planted the first orchards, usually olives. The complete Eurasian agricultural system with grains and livestock which has now diffused worldwide soon evolved roughly simultaneously in southwest Asia (the Near East), central Asia, and south Asia.[26]

Agricultural populations grew. Babies arrived more often than among hunter-gatherers. To feed everyone, farmers needed another source of energy. Humans could only hoe so much ground, plant so many seeds, and harvest so much grain. Slaves provided extra energy but added mouths to feed. Animals that ate grass, which needed no cultivation, offered a better solution to energy needs. Before 7,000 years ago, some innovative person first yoked oxen to pull a pointed stick across the ground—the first scratch plow. In following centuries, as cities and populations grew and needs increased, people harnessed the energy of draft animals for such hard, tedious jobs as milling, grinding, pressing, pulling heavy-laden carts, or lifting irrigation water (just as we still "harness" rivers, waterfalls, natural forces, or atomic energy for power).[27]

The muscle energy of animals presented a mixed blessing. When people tightly integrated plant and animal production and used animals to plow, haul, thresh, manure, and provide milk and meat, they made farming more efficient, productive, and healthier. Draft animals sped many processes and provided power to cultivate more fields. Their manure fertilized the land they worked and fed on. However, a pair of oxen plowed six to nine times as much as humans could hoe but gave humans six to nine times as much grain to weed and harvest. Labor saved in the field was frequently invested in caring for and feeding animals. Some arable land must be sacrificed for pasture. Fodder must be laid up for the winter. Space in barns, stalls, stables, or homes must be made for animals. Finally, sometimes farmers and their animals traded diseases and parasites, and passed them to their neighbors and which traders carried to distant lands.[28]

Independently across six continents, humans domesticated a huge array of plants and animals, from wheat to pineapples and from dogs to ducks. Few of these crops and animals stayed long in their native regions. As Jared Diamond and others have noted, Eurasia enjoyed an advantage over all other continents with its very long east–west axis that stayed mainly between the Tropic of Cancer and the Arctic Circle.[29] This vast temperate zone encompassed the most important cradles of agriculture, from New Guinea to Europe. The north–south axes of Africa and the Americas crossed climate zones and lacked easy routes of passage, while long east–west coastlines stretched from southeast Asia, to India, the Near East, Africa, and the Mediterranean basin and encouraged men to sail boats between most one civilization and another. The earliest civilizations were in communication. Trading ships plied the waters between Harappa, Sumer, and probably Africa.

Generally, plants and animals that thrived in one region also grew well in another. Traders and seagoing colonists spread new domesticates. Colonists took their plants and animals with them into Anatolia and the Balkans and around the Mediterranean basin, where natives often adopted them and sometimes domesticated local species as well. Imported animals replaced endemic island fauna, an early rehearsal of European "ecological imperialism" in America, Australia, and around the globe.[30] Innovations also quickly spread, which may explain the rapid expansion of domestication of draft animals throughout the Old World but not the New.

By 3,000 years ago, hunting and gathering had given way to farming and herding across the earth wherever agriculture and pastoralism would take root, completing the initial stages of anthropogenic environmental change.[31] Agriculture and pastoralism changed the earth and its climate far more than anything *Homo sapiens* had done before. Herders set fire to forests to make pasture. Other forests fell before polished stone axes and then more-effective metal ones. Exposed soils blew and washed away. Irrigated fields grew salty and barren. Herds overgrazed hillsides and exposed soil eroded. Thinner forest cover affected climate, beginning 8,000 years ago, as carbon in tree trunks escaped to the atmosphere as the greenhouse gas carbon dioxide. Increasing herds of cattle produced methane, an even more potent greenhouse gas, as William F. Ruddiman has noticed. Then 5,000 years ago East Asians began planting rice in diked and flooded fields. Weeds and rice straw rotting under water produced lots of methane. Methane from the spreading culture of rice derailed the natural climate cycle that would have turned colder and taken the world into another Ice Age. This mild global warming worked out well for humanity. Methane heating stabilized the climate, kept the earth warm and mostly ice-free, and fostered the rise of world civilizations.[32]

Civilizations, trade, and incipient capitalism

Agriculture grew more than food. Farming also raised population, cities, trade networks, and empires. Egyptians, Mesopotamians, and Harappans relied on annual floods to water and fertilize fields. To exploit the soil more fully, Mesopotamians and Egyptians expanded local irrigation systems into regional networks, which required administrative oversight—framework for a government. Harappans used only small-scale irrigation but exploited the monsoon seasons to grow two crops a year.[33]

Meanwhile, 7,000 years ago in present-day Bulgaria, far from these early civilizations, craftsmen made a fateful discovery. Using kilns designed for ceramics, they found that heating certain rocks from the Balkan and Strandzhe mountains yielded copper and gold.[34] Copper made superior tools and weapons that were easier to sharpen than flint. Gold never tarnished and gleamed and glittered delightfully, although too soft for most practical uses. Demand for these metals grew rapidly from farmers, craftsmen, warriors, and the wealthy looking for status markers. Metalworkers far and wide learned to smelt other metals and experimented with alloys. Bronze did not take long to appear, then iron and lead. Demand for metal products accelerated between 4,000 and 3,000 years ago and continued to rise until the fall of the Roman Empire.

The ancient human activity of trade grew more capitalistic and drove ever greater exploitation of environmental resources. The grassy fertile plains of Mesopotamia could not supply the needs of Sumer, the world's first civilization. Only extensive trade could do that. Wood was needed to fire bricks for building and for beams and furniture, and for fuel, tools, weapons, and vehicles. Shrines and temples required material for statues and structures. Clothing required flocks and pasture for wool or, in India, fields for cotton. Sumerians imported these items, along with flint and obsidian for tools and prestige goods like lapis lazuli, carnelian, and chlorite. Later came copper, tin, gold, and silver as well. To get them, urban traders traveled to the Zagros Mountains, Anatolia, Iran, Afghanistan, and the Persian Gulf.[35] Egypt and Harappa had similar needs and their merchants too traveled far in search of resources.

Here also rose the forerunners to the system of capital, banking, and trade that underlies today's global economy. Sumer's city-states learned the lesson that today enriches Amazon.com and Walmart, that profit goes to the middleman. Major cities designated open space for economic activities—the marketplace. They grew wealthy as they exploited price differentials on goods. Byblos, for example, grew fabulously rich as the main port through which exports of copper, silver, lapus lazuli, wine, oil, and cedar went to Egypt.[36] A merchant class arose and developed such basic business forms as partnerships, agents, and indebtedness. Large temple complexes grew up around shrines to important gods, where priests worked full-time to keep deities pleased and offerings piled up. Temples and retired traders acted as banks and lent this capital to merchant traders. Around 5,200 years ago, Sumerian accounting symbols evolved into the first writing system.[37]

Civilization in all its glory had arrived. On the fruited plains between the Tigris and Euphrates, the walls and gates of the first great cities towered above amber waves of grain. In southwest Asia, south Asia, east Asia, Africa, Europe, and the Americas, every agricultural society fostered cities. Between 5,200 and 4,200 years ago, the number of cities worldwide with more than 10,000 souls rose from a handful to more than two dozen.[38] The city was something truly new under the sun. It teemed with thousands of people who ate food that they did not have to raise themselves. Slaves, craftsmen, traders, merchants, priests, warriors, and rulers jostled in the streets. Rulers, priests, and merchants patronized artists and architects. After the invention of writing, Mesopotamian cities boasted the first archives, libraries, institutions of learning, literature, mathematics, geometry, astronomy, science, and philosophy.

Commercial networks tied together the civilizations of India, Mesopotamia, Egypt, the eastern Mediterranean, and, more tenuously, east and southeast Asia. Far from essential or desirable resources, Sumer and the city-states of lower Mesopotamia sought by means of conquest or colonization to control either the sources of goods or the routes they passed over. Outlying communities specialized economically to serve the needs of the cities. Gold and silver acquired standardized value and functioned as media of exchange. Isolated on the other side of mountains and deserts from India and the Middle East, China acquired metallurgy (and, possibly, the idea of writing) via central Asia trading routes and cultures.[39]

The glory concealed much darkness. Farming bred inequality. It is not possible for property in land and animals to be divided among villagers or heirs perfectly equally or fairly. Moreover, as a form of wealth, cattle multiplied social disparity to a degree unknown in civilizations that lacked draft animals, as in America. A farmer deprived of draft animals was doomed to poverty. The powerful divided crop surpluses inequitably or appropriated them altogether. Warriors could in effect extort the profits of farmers, who only reluctantly would abandon fields, crops, animals, irrigation works, and homes in which they had invested so much labor.[40] From then until now, the wealthy and mighty attract riches and power like filings to a magnet unless societies create mechanisms to control inequality.

Agricultural surplus fueled war and unfree labor. In service to the common good, rulers took in taxes and organized labor to maintain and construct irrigation works. But they also warred against their neighbors for profit, tribute, resources, or more tax revenue. Rulers and temples accumulated vast tracts of land and commanded the labor of many farmers. Labor levies and slaves built monuments and palaces in addition to public works.

Civilization and environment

Growth and trade took a mounting toll on the environment. Even on the eve of the Bronze Age 6,000 years ago, Near Eastern forests were in crisis. Trees were felled to allow farmers and herders to fill more bellies. More (and more prosperous) people needed beams for new or larger houses, palaces, and temples.[41] Extensive trade required timber for fleets of ships. Workers cut cedars and pines in Lebanon for wood for Egyptian tombs, ships, sarcophagi, columns, doors, and roofs, and for resins for mummies and medicine. Extraction and smelting of metals accelerated with rising trade in the middle of the third millennium BC. Pits and mines scarred the remote mountains where ores concentrate. Mine tailings leached arsenic and toxic chemicals into watercourses. Smelting and metalworking required great quantities of wood for heat, competing with the needs of cooking and firing bricks. (The contemporary city of Mohenjo-daro alone required at least five million fired bricks.) Miners both free and slave needed wood for housing and cooking fires, sometimes transported from a distance. Inefficient ancient smelting methods produced more pollution than all European copper smelters in the nineteenth century. The soil around smelters accumulated toxic metals and chemicals. Smoke from the smelters of antiquity blew around the globe and left distinctive layers thousands of miles away in northern European bogs and on Greenland glaciers.[42] Overexploitation exhausted ores in many locations.

Growing populations pushed agriculture into less-suitable land, where irrigation was necessary. Erosion from deforested and overgrazed mountains and hills silted up irrigation works. Irrigation brought salts to the surface as water evaporated in the high heat, which slowly made fields barren. Where annual floods did not refresh the land, crops drew nutrients from soils faster than manuring could replenish them.[43]

Prosperous cities attracted attention of empire-building conquerors. Empires made economies run more easily. They broke down barriers, built infrastructure and roads, protected travel, and promoted trade and prosperity. Empires also led to the metropolis, the nexus where trade and administration brought immense wealth through taxation, trade, tribute, and booty. Exchange for profit, which had been the practice only of merchant-capitalist traders, now spread down the social scale and affected daily life.

Empires rose during periods of climatic stability. Fair weather and rain promoted prosperity and masked injuries to the environment. Changes of temperature and rainfall patterns brought civilizations crashing down. About 4,200 years ago, shifts in atmospheric and ocean circulation caused severe two-century drought and climatic cooling. Chaos and crisis erupted in Greece, Egypt, the Levant, Mesopotamia, the Indus valley, and the Yangtze River valley.[44] Sumeria, on the salt-burdened soils of southern Mesopotamia, never recovered its prominence. Another cooling and drying event about 1,400 years later affected the eastern Mediterranean. Archeology records the burning of scores of cities. Ancient Greece entered its Dark Ages. The Hittite, Kassite, and Egyptian empires collapsed.

Money, commodities, and empire

A new economic and environmental era began with another innovation in metals in another mountainous region far from the centers of the great civilizations. Around 560 BC, King Croesus of Lydia in western Anatolia struck the first coins. Certain items of exchange with a standard value had long lubricated trade. However, trading in measures of barley, silver, and gold, or sometimes copper, bronze, livestock, cotton, and foodstuffs, always involved certain inconvenience or risk. Were there stones in the barley or base metals in the gold? Was the livestock healthy? Was the local shekel (a weight measure) the same as a shekel at home?

The usefulness of standardized coins, so obvious to us today, dawned on people only slowly. It took a couple of centuries, but all the ancient world accepted the value of coins—identical, trustworthy, interchangeable, hoardable, desirable. People now made most exchanges with gold, silver, and bronze coins and came to think of almost everything in terms of monetary cost. Household items, labor, energy, slaves, land, wealth, livestock, food, homes and other structures, events, travel, taxes, books, education—what does not have its price? Today, money rules all. Modern capitalism and, indeed, life as we know it are unimaginable without it.

In Greek hands, money transformed economics. The tip of the Balkan peninsula where the Greeks made their home lacked the fertile, well-watered soils of the great civilizations. But rich veins of silver ran through the mountains. Greek cities began minting coins. States with mints now played a large, central monetary and economic role. Rulers needed a stable monetary policy to prosper. Money simplified taxation and made it much easier to raise and maintain armies. It affected temples and religion. Money greased the wheels of commerce and transformed accounting, contracts, credit, and wages. Coins eased the path to comfort and riches for nonagricultural elites and constituted an easy means to accumulate capital or measure wealth.[45]

To profit from land with so little fertile soil, Iron-Age Greek city-states created a new form of capitalism that presaged the plantation capitalism of the European empires two millennia later. Farmers raised and processed high-value crops like olive oil and wine for export. Both were labor intensive, so farmers invested their capital in foreign slaves captured in war or bought from hard-luck foreign families. Slaves also did the

Figure 1. The Antimenes Painter, Amphora, 520 B.C. Two men and two youths, likely slaves, harvest olives. The Greeks and Romans made the Mediterranean Sea a highway for traffic in olive oil and wine, agricultural commodities produced by slaves on large estates. (British Museum 1837,0609.42)

unpleasant and dangerous mining of silver and salt. Commodity agriculture, little different from later plantation agriculture, was a new system that required significant capital investment, concentrated energy supplied by human and animal muscle, and processing facilities. Greek prosperity was also the gift of cheap transportation by sea, as wind-powered merchant fleets took the olive oil and wine to distant markets. This early plantation capitalism made the Greek economy hum and supported the brilliance of its art, architecture, science, and philosophy.

Greek merchant capitalists developed sophisticated single-entry bookkeeping but did not advance past the traditional merchant capitalist model of dissolving partnerships after each voyage. Greek plantation capitalism also did not make the transition to industrial capitalism. Perhaps the relative scarcity of wood and coal in the Mediterranean kept some inventive Greek—who could make such astonishingly complex precision devices as the Antikythera mechanism to calculate positions of celestial bodies—from devising a steam engine.

Greek population rose with Greek prosperity. Already in the sixth century BC, thriving city-states outran limited local agricultural capacity. Several cities founded colonies around the Aegean and Mediterranean. Grain imported from Egypt and elsewhere fed the Greeks and freed land for profitable commodity crops. Greeks discovered the great truth that trading societies importing their food avoid the ever-present peril of famine that agricultural societies face from failed harvests. At the same time, wealth did not widen inequality among free Greeks. The relatively small size of Greek city-states maintained relative equality among freemen. Cities also maintained relative equality because independent farmers and artisans filled the ranks of their constantly warring armies. As among white Virginians in the eighteenth century, Greek freedom and relative equality rested on the backs of slaves.[46]

Romans would model their huge estates, the *latifundia*, on Greek plantation capitalism. They built the most prosperous, most urban empire anywhere in history prior to the nineteenth century. The Roman Empire grew fat by connecting and protecting trade from Egypt to England and from Iberia to Armenia. Swarms of traders extended the reach of the Roman economy as they went to and fro across Africa and over the Indian Ocean. Excellent roads constructed to move troops around the empire became highways of commerce.[47] All Mediterranean lands benefited from the tideless and relatively stormless sea they bordered. They harnessed the energy of the winds to travel over its surface and to harvest its fish. In an early use of fossil energy, British Romans also used coal from all major British coalfields under Roman jurisdiction to heat public baths, smelt metals, warm homes, and use for a multitude of other purposes.[48]

Empire and environment

Roman power and glory peaked during four centuries of uniquely favorable environmental conditions called the Roman Climate Optimum. During this era, the Mediterranean received appreciable rain in summer, which no longer happens. At the height of the optimum, the sun smiled on the Roman, Parthian, Han, and Maurya empires in the Old World and Teotihuacan in the New. Dozens of world cities boasted more than 100,000 inhabitants. The tremendous metropolis of Rome peaked above a million while the Roman Empire counted about seventy million souls.[49]

Such a large, prosperous population taxed the environment hard. The vast number of cooking and baking fires, oil lamps, kilns, metalworking fires, and furnaces for public baths in large cities put urban dwellers under perpetual clouds of pollution. For supplies of wholesome clean water, the Roman Empire built hundreds of aqueducts (eleven for the city of Rome alone), often for considerable distances (160 miles for the Valens Aqueduct of Constantinople), and reworked watersheds. In the days before public transit, people lived crowded close together. The excrement and garbage of so many people and animals offered a huge challenge to authorities. Vermin ran rampant. City diseases stunted people's growth. The daily demand for food and fuel strained resources for long distances around. Axes were busy in forests everywhere to make room for farms and pastures, supply wood for fires and furnaces, provide building material, and furnish timber for the empire's fleets and merchant marine. The most likely reason Roman Britons used coal and often hauled it significant distances had less to do with any superiority of coal as a fuel and more with shortages of wood.[50]

The Roman Climate Optimum ended around AD 150. Extensive deforestation of the Mediterranean may very well have helped cause a shift in rainfall patterns. In any event, it exacerbated the problem. Disruptions to trade and agriculture culminated in the Antonine Plague, likely the first major outbreak of smallpox, which swept through Rome's dense cities and killed millions. The *latifundium*, plantation capitalism in its first version, vanished. In China, the Han dynasty fell.

Rome carried on at a reduced level, with a bit of a comeback in the fourth century. Then in the sixth century the climate turned cool and wet and brought catastrophe. Changing weather pushed Huns, Germans, and Slavs south and west. Western Rome fell, its population collapsed, trade faltered, and cities faded away. Coinage vanished from vast stretches of Rome's former territory. The eastern Roman Empire suffered the Justinian Plague, the greatest mortality event in history to that time, which left it weakened. Deforestation caused extensive erosion, especially in the eastern Mediterranean. The fabled harbors of Ephesus and many other places silted up, leaving malarial marshes that depopulated much of the coastline. By the end of the sixth century, Rome contained only perhaps 20,000 inhabitants.[51] Farther east, in India, the great Gupta empire disintegrated.

Considering the power and glory of Rome, the fall of the Roman Empire and Roman civilization shocks the imagination. Densely populated areas returned to forest, not to be cleared again until the ninth century or, in some cases, the nineteenth. Cereals grew poorly in these wet, cold centuries. Fields reverted to pasture. Herding replaced farming as the main source of food in the depopulated early Middle Ages. The ancient landscape was gone. A medieval one grew up amid Roman ruins.[52]

Climate changes that are disastrous for one people often bring good fortune to others. Central Asian pastoral populations thrived in the cooler, wetter weather of the late first millennium. Iraq prospered from the unusually wet climate. Newly converted Arab Muslims pushed back the weakened Byzantine empire. Between the eighth and eleventh centuries, the Islamic Levant flourished. Arab middlemen grew rich from trade and developed such improved business practices as an early version of modern double-entry bookkeeping.[53]

During this time, Europe slowly repopulated. Free land grew scarce. Farmers lost independence and freedom. Weak and fragmented central governments could not restrain local lords from forcing peasants into tenantry. Serfdom spread, although outright slavery diminished. Frequent warfare concentrated settlements in the shadow of castle walls or behind city walls. European civilization grew decentralized, insecure, unfree, and relatively impoverished.

In contrast, in China the population grew quickly with the spread of rice culture. Since peasants are easier to tax than lords, emperors restrained the power of local landholders. Taxes were kept low to forestall social unrest. A strong central government upheld the peace, which also allowed the decentralized population to live secure in farms and villages. Craftsmen and artisans lived in the villages, not in cities. Without strong guilds to interfere, innovations and advances spread quickly throughout the empire.[54]

What a long, strange trip it had been for *Homo* in all its varieties. When evolution (or aliens) bestowed extraordinary abilities to communicate, cooperate, organize, and improve technique, the earth became practically unrecognizable within a relatively short time. Humans made themselves such effective top predators that they caused extinctions wherever they took up residence. Their mastery of fire changed landscapes. They transplanted and manipulated food plants, mined flint and red ocher, and traded with their neighbors. Well fed, they increased, multiplied, and spread across the globe. People further altered ecosystems through herding and farming to support their growing numbers more efficiently. Populations grew, towns transformed to cities, capital accumulated, inequality and unfreedom increased, and from competition and violence kingdoms and empires rose.

Greek and Roman innovations accelerated the economy. Merchants took advantage of free wind energy to crisscross the Mediterranean Sea. Coinage sped and democratized exchange, created markets in slaves and commodities, and facilitated taxation, wage labor, and warfare. Slaves working for the profit of landlords on large estates produced wine and olive oil that sailing ships took across the seas, carrying grain in return. This early capitalist system supported a large and prosperous urban population. It also led to deforestation, species extinctions, water and air pollution, erosion, and climate change, just as earlier civilizations had caused salinization and soil exhaustion. Demand for metals created environmental "sacrifice zones" around mines and smelters. Plagues and changing climate toppled the Roman Empire. By the sixth century, this experiment in capitalism and the environment fell with a resounding crash.

Plantation capitalism would slowly rise again among Mediterranean nations over half a millennium later. This time, however, advances in ship and sail design turned the oceans of the world into Europe's Mediterranean. European capitalism and empires would show up on every land in the world. Resistance would be futile.

TWO

Trade and Empire

Columbus's accidental discovery

Christopher Columbus always oversold everything. Perhaps some unshakable sense of inferiority in a class-conscious society drove this son of a Genoese weaver to bravado and ardent ambition. He bluffed his way to winning the jackpot when his unrealistic claims set the events in motion that led him to America. His bold, boastful accounts electrified Europe and started a train of unforeseen consequences. Thousands of grasping, ambitious men swarmed across the oceans to make their fortunes, often heedless of or indifferent to consequences for strange people and strange lands far from the restraining hand of custom and authority. Millions died of disease and war. Millions more labored, suffered, and died in bondage. Relentless, profound environmental changes followed the paths of trade and settlement. European empires of mercantile capitalism and plantation capitalism encircled the globe and channeled immense profits and power to formerly marginal nations. Almost three centuries later, they fed the rise of industrial capitalism. Columbus foresaw none of this, but he made it all possible.

This all came to pass because this weaver's son hoped to win glory and renown by finding an easier route to China. The self-taught Columbus devised a novel plan. He avidly read Marco Polo's account of his travels to the fabled court of Kublai Khan. He corresponded with Florentine mathematician Paolo dal Pozzo Toscanelli, proponent of a theory that a ship sailing west across the Atlantic could reach the Far East. Fame, honor, and incalculable profit would surely redound to anyone who stood again before the Great Khan in his golden-roofed palace. Columbus desired obsessively to be the first to make that crossing and be that person.

Strong, at six feet well above average height, fair-haired and fair-skinned, self-confident, and determined, Columbus made a formidable salesman who could win the confidence of skeptical monarchs. He used connections from a fortunate marriage to present his proposal in 1484 to King John II of Portugal and in 1486 to King Ferdinand of Aragon and Queen Isabella of Castile of Spain. But in his eagerness to clinch the sale, he exaggerated potential profits. He painted a tempting, glittering fantasy of "great lands, . . . all very prosperous, rich in gold and silver, pearls and precious stones, and an infinite number of people." Columbus made audacious demands in case of success: ennoblement; the hereditary title of Dom; the rank Admiral of the Ocean Sea; appointment as perpetual viceroy and governor of all he might discover; a tenth of the king's share of "all the . . . goods bought, exchanged, discovered, or acquired from the region of his admiralty"; and the right to invest an eighth in future expeditions and receive an eighth of the profits.[1]

Columbus's sales pitch had its effect, but not immediately. John put him off while he secretly tried Columbus's plan and sent out a small fleet from the Azores, which at that latitude ran into the unfavorable trade winds and returned. The king altogether lost interest in risky adventures to the west after 1488, when Bartholomew Diaz passed the Cape of Good Hope, which offered a surer route to the Indies. Isabella found Columbus's pitch more tempting. She and Ferdinand had incurred a huge debt to conquer Granada and needed revenue.[2] Castile had concluded a peace treaty with Portugal that had given the Portuguese a monopoly on lands south and west of the Canaries, where lucrative African gold and slaves were found. By heading straight west from the Canaries, Columbus could sail straight through a loophole in the treaty to the Indies and all its riches.[3] After he threatened to take his proposal to the rival French, Isabella consented. Finally, in 1492, Columbus sailed the ocean blue.

Columbus must have been enormously gratified to find islands exactly where he expected them—and colossally disappointed to discover no rich civilization with golden roofs and silk-clad courtiers. Tropical islands with naked inhabitants without significant trade or wealth would not impress investors. Columbus now had to convince his backers to fund another expedition to Japan and China, which, judging by fields of cotton that he saw in what he called the Indies, must surely lie close by.[4] Consequently, he again vastly oversold his discoveries in his report to Ferdinand and Isabella. The numberless population of *La Isla Española* (the Spanish Island, or *Hispaniola*, as he called it), he claimed, was gentle, peaceful, and innocently naked, ready to be put to work and made Christian.

Soil fertility was "limitless." Harbors were "beyond comparison." Spices, cotton, mastic, aloe wood, and slaves abounded. Cuba was larger than England and Scotland and its mountains surpassed those of the Canaries. The length of Hispaniola's coastline exceeded Spain's. Inland regions boasted "great mines of gold and other metals." Columbus founded a town, La Navidad, "in the best position for the mines of gold . . . and for [trade] with the mainland . . . belonging to the Grand Khan." Whether from delusion or mendacity or savvy salesmanship, little of Columbus's report was reliable. Much was utterly false.[5]

Columbus paid a price for overselling his discoveries. Printed versions of his letter spread visions of Columbian Fantasyland across Europe. Spain burst with popular excitement. His second voyage to the Americas attracted plenty of investors and settlers. Columbus commanded a huge fleet of seventeen vessels laden with twelve hundred colonists, along with horses, cattle, pigs, goats, sheep, wheat and garden seeds, and fruit trees—a veritable ecological invasion.[6] Columbus sincerely believed that Hispaniola could be effectively colonized, if not as easily or profitably as he represented, but he lacked experience and skill to command such a large expedition. Colonists were all men (and no women) of the type who would abandon home and kin for a chance at easy wealth. On Hispaniola, when they found no pliable workforce and little gold, Columbus could not control them. He used violence, cruelty, and terror both to discipline his colonists and to force the native inhabitants to bring him gold.

When ships returned from Hispaniola to Spain for supplies, Columbus needed somehow to make the colony profitable. The islanders did not easily fit his understanding of the world's peoples. They were not Christian, so was it not likely they were enslavable Muslims? He rounded up about 500 islanders to send back as slaves and gave 650 more to settlers. He requested that the ships on their return to Hispaniola pick up molasses and sugarcane from the Portuguese island of Madeira. On his third voyage, rebellious colonists forced him to grant them *encomiendas*, a kind of feudal estate with the right to command tribute and regular labor from inhabitants. Frustrated with governing colonists and "Indians," Columbus soon headed off to do what he really ached to do: explore in search of the rich lands of the Great Khan.[7]

Columbus's exaggerated salesmanship succeeded in getting funding and ships but also raised expectations far beyond what he could deliver.

Angry colonists sent him back to Spain in chains. Ultimate consequences were far worse. Epidemics of Old-World diseases would surely have sent staggering numbers of Americans to the grave no matter which European first crossed the ocean. But European brutality and ill-treatment of Americans in the quest for wealth and profit began with Columbus and his Spanish colonists. Columbus's fixation on precious metals and his introduction of slavery and sugar ominously presaged the kind of empires that would soon grow in America and alter world history, although they would ultimately benefit Spain's enemies more than Spain. In those countries, not in Spain, lay the future development of capitalism. But all European empires contributed to massive environmental change.

Genoese, the capitalists of empire

Columbus was an incompetent governor, but he acted according to a logic typical among Genoese of his day. Genoa and other northern Italian maritime republics had revived practices with roots deep in Mediterranean history and grown wealthy from trading empires, colonies, gold, sugar, and slavery. Genoese engagement in setting up the Iberian empires, beginning well before Columbus's voyage, taught lessons that the Portuguese, Spanish, then Dutch, French, and English learned well. Genoese manufacturing, banking, and textile industries also foreshadowed Dutch and British developments. Genoa formed the bridge from the early capitalism of the Romans to the imperial, plantation capitalism of the great world empires. Italians and Genoese like Columbus planted Italian capitalism in growing world empires of countries to which its methods were foreign.

Genoa, Venice, and Florence were the capitalist superpowers of the age. The cities of northern Italy alone among urbanized regions had survived the fall of the western Roman Empire. The Crusades resuscitated withered Mediterranean trade networks and put Italy on the road to prosperity. Soon, Italian merchant and naval fleets ruled the waves of the Mediterranean and Black Seas. Rowers' muscle power and free wind energy sped galleys full of spices, cottons, and silk from the east to Italian ports for European customers in the West. Silver from the mines of Central Europe and fine cloth from Flanders crossed Alpine passes to Venice, while Flemish and French textiles and commodities funded by Italian bankers arrived in Genoa. Gold dust from Ghana left north

African ports which Venetians minted into ducats and Florentines into florins. Italian ships returned to eastern markets bearing cash, textiles, and other goods.

Italians made important innovations that propelled ancient mercantile capitalism toward modern business practice. Historically, every trading partnership had dissolved after each venture ended. Italians set up the first permanent family based firms, the ancestors of modern corporations. Managing far-flung trade networks required trustworthy agents. Family firms could place reliable relatives in foreign ports. Italians moreover pioneered many modern business practices, including double-entry bookkeeping, debt instruments, and banking.[8]

The Venetian and Genoese republics also created the model for overseas colonization and imperial expansion. They established colonies and urban merchant enclaves operating under Venetian or Genoese law. Venice planted colonies around the Adriatic and in the central Aegean and eastern Mediterranean. Genoa privatized its empire. Its leading families created and ruled colonies around the northern Aegean, Black, and western Mediterranean Seas. Privatized colonization by Dutch, English, and French companies would follow Genoese precedent.

The Dutch and British would also mirror the way northern Italy turned manufacturing zone. No mere entrepôts, Italian cities imported raw materials and exported textiles, glass, and other goods. They imported raw cotton from the Levant and exported dyed cloth. Florence, Lucca, Milan, and Genoa finished Chinese silk brought in through Levantine ports. With wool from Spain, north Africa, and England, Florence produced some of Europe's finest cloth.[9] A third of the fifteenth-century Florentine population worked in the wool trade, as did a third of Venice's a century later, while about a third of Genoa's workers labored at silk, wool, and cotton production.[10] Merchant-entrepreneurs like Columbus's father employed eight or nine workers and paid piecework to many more, each of whom worked in a shop or their own home to perform one step of the many that turned raw wool into beautifully colored fine fabric, a division of labor much like the one for making pins that Adam Smith would so famously admire.[11] For the first time in Western history, more than half the population of a city, along with a substantial proportion of nearby towns, worked for wages—a proletariat.[12] Industry also mechanized. Centuries before Lancashire's cotton mills, Lucchesi,

Bolognese, and Milanese mechanics devised water-powered mills that employed hundreds of women under severe discipline to throw and wind silk thread. Hundreds of mills, some extremely large, spread across the hills and high plains of northern Italy.[13]

Fatefully, the Italians also bought and sold people. Genoa, especially, and other Italian cities participated enthusiastically in the slave trade, which to Columbus must have seemed quite unexceptional. European slavery had almost become extinct—except in the Mediterranean.[14] The Catholic Church limited enslavement to non-Christians who resisted conversion, so slave raiders used any convenient pretext to take non-Christians. Enslavement was common in interreligious wars. Christian and Muslim corsairs and pirates raided each other's coasts and shipping, holding captives for ransom or selling them into slavery. Between 1450

Figure 2. Stefano della Bella, Loading a Galley, from *Views of the Port of Livorno*, 1654–1655. Renaissance Italian capitalism in action: workers load a galley with trade goods in Livorno. The etching hints at slavery's ties to religion, capitalism, and war. At the far right, a man, possibly a slave, is whipped. At left is a portion of the Monument of the Four Moors commemorating Ferdinand I of Tuscany's victory over the Ottomans. Four slaves in chains surround the pedestal, one of whom, visible here, has black African features. (Metropolitan Museum of Art 1986.1180.574. Bequest of Grace M. Pugh, 1985)

and 1850, at least three million people from around the Mediterranean found themselves in miserable slavery.[15] By 1400, slaves made up perhaps ten percent of the populations of Genoa and Venice, the highest proportion in Europe. Most slaves came from Italian colonies around the Black Sea (*slave* comes from *Slav*). The rest were Muslims, with a few from sub-Saharan Africa. Slavery was harsh and sexual abuse rife. Recalcitrant or runaway slaves faced brutal punishment. Slaves were slaves for life, as were any children, although manumission to free a person was not uncommon.[16]

Sugar and slaves

One signal Italian innovation was the sugar plantation, close kin to the *latifundium* and ominous ancestor of the lucrative American slave plantation. Sugar had only sweetened European cuisine for two or three centuries before Columbus sailed. Ancient Romans had known sugar as a costly medicine from India. Europeans of the early Middle Ages knew it not at all. In 1220, the first ship sailed into an English port with this luxury item in its cargo. In 1277, Genoese convoys laden with sugar and spices began regular trips into the Atlantic bound for Southampton and Bruges.[17] Sugar signified high status and sophisticated taste, so nobles' cooks put it into almost every dish, sweet or savory.[18]

Sugar had arrived in the Mediterranean with Islam. The spread of Islam out of Arabia and into Persia, India, the Levant, northern Africa, and Spain in effect created a giant trade zone from India to the Atlantic. From China and India, Muslims brought sorghum, citrus fruits, rice, cotton, and sugarcane, which enriched cuisines throughout the Islamic Mediterranean. Irrigation techniques spread as well and enabled larger harvests and denser populations. When conquering Crusaders arrived with their own crops and agricultural techniques, Muslims left or were expelled. Irrigation works deteriorated. The new crops disappeared—except for sugarcane.

At the nexus of capital, regimented labor, international trade, and consumerism stood sugar. Cane turns into valuable molasses or sugar only with plenty of water, much labor, and expensive processing works. Sugarcane thrives and gets sweetest in the hot, wet tropics. The Mediterranean region, at the northern limit of the plant's range, gets no summer rain, so Arabs raised it on prime irrigated land along rivers and in marshland.[19] With grueling labor, farmers harvested cane taller than themselves, removed the leaves, and brought the cane to presses quickly before it spoiled. Workers then cut it into short pieces, crushed them, and boiled the juice. To get a more valuable commodity, they added water back to the juice or crystals and reboiled it as often as necessary to get lighter color and more-profitable quality. Energy from horses, oxen, or water powered cane presses. Heat energy from wood boiled the juice. Human muscle-energy did everything else. Peasants and tenants grew cane as a side crop near palaces with processing facilities or near waterways for easy shipping. Muslims did not use slaves for agricultural production, although large sugar estates with occasional forced peasant labor appeared later in Egypt, and Moroccan sugar-growers probably used sub-Saharan slaves by the sixteenth century.[20] The elite consumed some of the sugar. They sold most of it very profitably to Italian or Jewish merchants to carry to European buyers.[21]

Christians began raising cane in the eleventh century in the Crusader states, for which sugar provided needed revenue. After the Crusader states fell in 1291, the last King of Jerusalem and his knights fled to Cyprus with sugarcane in their luggage. There they created feudal demesnes on the model of those they had known in Europe. Serfs and tenants raised cane. Italian merchants bought the sugar for customers in western Europe.[22]

This arrangement did not last long. Thriving civilizations around the earth felt a literal chill at the turn of the fourteenth century as the Medieval Warm Period yielded to the Little Ice Age. Warm summers turned cold and wet. Crops failed. On the other side of the globe, changing climate ended the high Mississippian and Pueblo cultures in North America and Tiwanaku empire in the Andes. Famine stalked Europe. Social, political, and religious chaos spread. The civilization of the High Middle Ages tottered and fell. A plague swept out of the east. Crossing the Mediterranean on Italian merchant ships, the Black Death stepped ashore in Europe in 1347 with scythe swinging and for four years mowed down the living.[23]

Cyprus's population plunged. Landowners faced labor shortages and bankruptcy. Genoa's merchants spied a business opportunity in emptied villages. Their ships sailed into Cypriot ports with Greek, Bulgarian,

Muslim, and Tatar slaves, mainly from the Genoese Black Sea colonies. Cypriot landowners expanded sugarcane production with slave labor and grew wealthy. Cypriot sugar acquired a reputation as the best in the world. Genoese firms aggressively leveraged control of banking and the slave and sugar trades to all but rule Cyprus by the end of the fourteenth century.[24] Genoa was now deep in the sugar and slave business and the stage was set for tragedy to play out around the Atlantic.

Environmental costs of Italian prosperity

The trade and manufacturing of the Italian maritime republics, which at the peak of their Renaissance glory were the richest cities in the world, sent ripples of environmental impact huge distances in all directions, a foretaste of the environmental impacts of colonial world empire and of industrialization. Forests thinned or fell back. Commercial crops exhausted soils. Digging or mining stone and ores left holes on the surface or deep underground. Mining and industrial processes poisoned earth, water, and people.

Close by were the forests. As in the ancient world, economic growth demanded a lot from scrubby, thin Mediterranean forests, never as thick as cooler, wetter northern forests. Venice alone needed thousands of oak pilings to keep itself from sinking in the soft soil of its island in the lagoon. Italian cities, like their competitors and enemies, needed tremendous wood resources for navies and merchant marines.

Italy's large textile industry put pressure on forests as well. Demand for wool encouraged expansion of sheep herds. To produce England's most valuable export, wool, for Flemish and Italian clothmakers, enclosure of common lands replaced traditional peasant land-use with sheep pastures. In Spain, Christians reconquered Iberia from Muslims in 1492 and transformed lush, irrigated valleys into less-productive farms, or pastures for vast herds of sheep and cattle to supply Italian textile and leather industries.[25] Southwest Spain suited cattle best. The rest of Spain was sheep country. More than three million sheep grazed Spanish pastures by the sixteenth century. Shepherds cut or burned forests for sheep pasture and caused significant deforestation. So many sharp hooves churning the soil and so many mouths devouring grass in the warm, dry Spanish climate impoverished and transformed the land.[26]

Wood was also fuel for boiling sugarcane juice into molasses or sugar. Cyprus and Sicily, the two largest islands involved in sugarcane production, seem to have been sufficiently large to provide fuel without exhausting their forests. Rising prosperity and land exploitation certainly put stress on soil and forests. Since sugar processing used low-grade fuel wood and woody brush, not mature timber, Cyprus's Troodos Mountains remained well-wooded. Nevertheless, restricted fuel supplies limited mills to the first steps of sugar processing. Sugar works in Venice and Genoa with better access to fuel refined the sugar to the most profitable quality.[27]

Many other manufacturing processes required wood fuel, notably glassworks, which also required a range of raw materials. Venice, Verona, Padua, and a few other cities produced the world's best glass, which found customers from China to America and Africa.[28] Glassmaking was technically challenging, logistically complicated, capital intensive, ecologically demanding—and highly profitable. A classic capitalist endeavor, a good glassmaking facility required a large investment in facilities and raw materials and employed about twenty workers. Uniquely among glassmaking centers, Venice had to import all raw materials. To reach 1,150° to 1,200° C. (about 2,000° F.) to melt glass, its furnaces burned about 2,200 cords (3.3 million board-feet, or 7,800 cubic meters) of alder or willow wood in a year, which was floated down the coast from the forests near Cervignano. For silica, water mills crushed pebbles from the Ticino and Adige Rivers 120 miles away. For flux, Venice each year imported 10,000 sacks of ash of a plant found in Syria, Egypt, and Libya. Manganese to make glass clear came from mines in the Piedmont northwest of Milan. Copper, silver, lead, and tin gave color to glass. Crucible clay came from Valenza, south of Milan, and from Constantinople. Stone slabs for furnaces and chimneys were hauled 45 miles from Vicenza.[29]

Farther away, gold and silver mining was especially environmentally destructive. A perpetual European trade deficit drained the continent of currency. To mint the coins to buy eastern commodities, Italians imported silver and gold. Silver came from deep mines in Tyrol, Bohemia, Silesia, and the Balkans. Rain and snow leached toxic chemicals from tailings and washed them into streams and rivers. Large numbers of sturdy timbers kept mineshafts from collapsing. To keep mines from flooding, miners constructed elaborate wooden machinery powered by animals or water, which often required rerouting water courses. Wood and charcoal fired the furnaces that smelted silver from ores. Forests retreated from the mouth of every mine and, as local timber was exhausted, along easy transportation routes like rivers. Deforestation around the mines of central Europe alarmed many, yet not until the eighteenth century did Germans devise conservation measures.[30]

Removing silver from the metals with which it occurred was a highly toxic process. German miners developed two effective techniques to separate out silver. In the fifteenth century, they (or some alchemist) discovered liquation, in which workers repeatedly melt ore with lead to separate silver out of copper. The rapid spread of this new process increased demand for lead, a dangerous toxin that smelting sent into the air to be absorbed by soils and living things. A century later, German miners started using mercury, which is extremely toxic, as an amalgam to separate silver, a simple, easy, and exceedingly hazardous technique. Large mercury mines opened in Almaden in Spain and Idrija in modern Slovenia to provide mercury for silver mines in Europe and America. Mining mercury-laden cinnabar ore poisoned workers, their neighbors, and the environment. In the final step of the amalgam method, workers boil off the mercury, leaving silver. Mercury vapor poisons everyone and everything nearby and downwind and lingers in the environment around smelters for centuries after mines play out.[31]

Gold mining in Africa caused more environmental mischief. Kings, chiefs, and local elites generally used slaves and forced or hired labor for mining. Local people and migrants attracted to the goldfields also often mined for gold, paying hefty royalties to elites for the privilege. Neither slave nor free miners worried much about environmental damage. Workers dug sand or gravel out of streambeds or from the bottom of open pits from ten to 80 feet deep. They rarely restored abandoned diggings. Mining required tremendous amounts of water, often from diverted waterways, with which to wash away sand and gravel and leave heavier gold. The technique silted streams, killed or drove away fish, and made water unpotable for downstream villagers. To a degree, custom and religious prohibitions protected land and water resources for local communities, although not for those downstream.[32]

Even the spice trade affected the environment four thousand miles away. Demand in China, India, the Levant, and Europe for spices was high, but the supply of cloves, cinnamon, and nutmeg was inelastic. Local people harvested them from wild trees that took too many years to get established to cultivate. The range of black pepper, on the other hand, easily expanded to meet demand. Cultivators brought this tropical vine out of its native Malabar and planted it in rainforests along the Indian coast and in Sumatra and Java. Pepper, like most other commodities in the holds of Italian galleys, enriched landowners, rulers, and middlemen, not the sharecroppers and workers who toiled to produce it.[33]

Genoese in Iberian exploration and empire

Columbus, sugar, and slaves might never have come to Hispaniola had it not been for the War of Chioggia in 1381. Venice defeated Genoa, locked up the spice routes, took over the Cyprus sugar trade, and made the island an outright colony in 1489. Then, in 1453, Ottoman Turks took Constantinople, extinguished the ancient Byzantine empire, and closed the Bosporus to Christians. Genoa could no longer reach its Black Sea colonies.

Genoa scrambled to recover from these calamities. The republic looked west for opportunities to replace lost trade and slaves. The strong nations of the western Mediterranean could not be colonized. The Genoese had to operate within others' empires. They established trading communities in Seville, Lisbon, Porto, La Rochelle, Southampton, and Bruges as well as Salé and Safi in Morocco. The Spanish found Genoese aggressive, opportunistic, and greedy.[34] A mayor of Seville complained to the king that he was just "not unscrupulous enough to deal with them."[35] But Genoese enterprise and banking shaped economies all over the western Mediterranean and western Europe.

Genoa now got sugar from Aragonese Sicily and possibly Corsica.[36] Aragon also brought sugarcane (without agricultural slavery) from Sicily to Valencia and Portugal established a sugar industry in the Algarve, while Genoese merchants controlled financing and trade.[37] The Genoese found new sources of slaves in the Canaries and along the African coast and sold Muslim, Berber, and black sub-Saharan slaves in Iberia. By the late fifteenth century, slaves made up over ten percent of the population of Lisbon, Seville, Barcelona, and Valencia and remained common in southern coastal Portugal and Spain for another century or two.[38]

Young Columbus was one of these opportunistic expatriate Genoese. In 1476, he was in a Genoese convoy making its way to London and Flanders when enemy ships sank the galley he was on off the Portuguese coast. Grabbing an oar, he swam to shore and made his way to Lisbon. There, the Genoese Centurione family welcomed him and made him an agent. He soon was aboard a Genoese ship headed to London and Flanders. With Lisbon as his new base, he voyaged to Bristol, possibly Galway, maybe Iceland, and to Madeira and Guinea.[39]

Columbus followed the path of many Genoese sailing in the service of Iberian monarchs, who needed their experience and capital. The Genoese navigator Lancelotto Malocello discovered the Canaries in 1312 for King Denis of Portugal. The Portuguese navy was founded in 1317 when Denis made Emanuele Pessagno (or Manuel Pessanha) hereditary admiral if he would provide twenty Genoese galley captains. Denis's son Alphonso IV sent a mapping expedition to the Canaries in 1341, commanded by Florentine Angiolino del Tegghia de Corbizzi and Genoese Nicoloso da Recco. Another Genoese, Antoniotto Usodimare, discovered the Cape Verde Islands in 1456. When Henry VII of England decided to join the race to find a westward route to Asia, he sent Genoa-born Giovanni Caboto (John Cabot) on voyages of discovery in 1496 and 1498.[40]

Italians in Spain and Portugal provided not only the mariners but much of the funding for Iberian exploration and empire (the Fugger family of German silver mining magnates being the other major source of loans). Italians helped to fund Columbus's voyages. Florentine banker Giannotto Berardi and Sevillian Genoese merchants loaned Columbus the money he was required to invest in his first voyage. Two of the merchants, Francesco Pinelli and Francesco Rivaroli, were at the time also financing the Spanish conquest of the Canaries. Pinelli, whose family invested in the first and second voyages, took part in both voyages and remained on Hispaniola after Columbus returned to Spain. Four Genoese merchants provided the cargo for the second voyage. The Centuriones advanced him cash for the third. Rivaroli loaned him money for his fourth and last voyage, which carried eight Genoese, including the captain of one of the four ships.[41]

Genoese kickstarted the Atlantic slave trade. In February 1494, Columbus sent twelve slaves from America to his creditors Pinelli and Berardi. Pinelli, Berardi, and Rivaroli were slave merchants who sold

Guanche slaves from the Canaries and Muslim slaves taken during the 1487 siege of Malaga (whose surrender and enslavement Columbus had witnessed). In 1494, Columbus's brother Bartolomeo formed a partnership with Berardi to sell American slaves. When Berardi died soon after, his executor, Florentine Amerigo Vespucci, sailed from Seville in 1499 and brought back two hundred American slaves. However, Columbus's shipment to Spain in 1495 of three hundred slaves (survivors of the 500 he sent) disturbed Queen Isabella, who regarded Americans as subjects and potential Christians. She banned enslavement of American Indians. Pinelli's descendants abandoned traffic in American slaves and invested instead in the African slave trade.

With the enslaved Indians and other failures, Columbus fell precipitously from grace in 1500. Master salesman as ever, he managed to plead his way to commanding his last exploratory voyage to America. Discovery of gold in Hispaniola completed Columbus's rehabilitation and he died a wealthy man.[42]

Technology, the quest for gold, and beginnings of empire

The entire European commercial and colonial enterprise in America, Africa, Asia, and the world was made possible by advances in transportation technology, capitalism's indispensable attendant. Columbus would never have reached America with Genoese galleys and navigation skills. Galleys could not tack close against the wind. Medieval navigators had few ways aside from stars and dead reckoning to steer a course. The Genoese dared not venture farther beyond Gibraltar than Morocco and England. Had ship design and navigation skills been up to sailing on the open ocean, they would have gone much farther. In 1291, two Vivaldo brothers set off from Genoa in well-provisioned galleys to sail around Africa to India. They were never heard from again. Inability to sail against wind and current or to navigate open ocean may well have prevented the return of the expedition, had it reached as far as the Gulf of Guinea.

The Portuguese, not Italians, took the first steps toward fast, reliable ocean transportation. The Iberian nations yearned to retake Muslim land for Christianity, especially the remaining Islamic states in Spain and the conquered lands of the threatening Turks. The impetus for a new ship design came after 1415, when Portuguese conquered Ceuta, Morocco, on the Strait of Gibraltar, a port for gold caravans from Ghana.[43] When, to Portuguese dismay, the caravans now headed instead to Tangier, they decided to reach the gold regions by sea.[44]

Exploring ocean routes to distant gold sources required ocean-worthy ships. The Portuguese came up with the caravel, the first ocean vessel that could tack closely against the wind. To carry cargo and supplies for long voyages, they modified the caravel as the more capacious carrack. To fend off or attack pirates or enemies, carracks carried plenty of cannon. Raised "castles" fore and aft allowed defenders to rake hostile boarders with musket or cannon fire. From the carrack evolved the powerful galleon at the turn of the sixteenth century. Other nations soon copied the

Figure 3. Pieter Bruegel the Elder, Three Caravels in a Rising Squall with Arion on a Dolphin, from *The Sailing Vessels*, 1561–1565. The caravel and its larger cousin the carrack revolutionized ocean-going trade and exploration. The caravel's keel, square rigging, and triangular lateen sails made it remarkably maneuverable. Its raised decks, called "castles," allowed defense against borders and its cannons made it formidable. (Metropolitan Museum of Art 59.534.24. Bequest of Alexandrine Sinsheimer, 1958)

Portuguese designs and made improvements of their own. On his first voyage in 1492, Columbus commanded a carrack and two caravels.

Advancing navigation techniques allowed the Portuguese to sail confidently far out to sea to catch the most favorable winds and currents. The dry compass, mariner's astrolabe, quadrant, cross-staff, and improved mapping (by skilled cartographers like Columbus's brother, a mapmaker in Lisbon) allowed voyages out of sight of land in any weather and south of the equator, beyond sight of the North Star.[45]

The consequences were nothing short of revolutionary. Portuguese could sail almost anywhere faster and more reliably than any other seagoing nation. Ship after ship ventured south along the African coast, catching tuna and sardines to finance exploration and picking up trade along the African coast, which beyond the Sahara virtually no native African or other traders sailed.[46] In 1471, explorers at last reached the rivers leading from gold-mining areas and captured the gold trade from Sahara caravans. To protect this hugely lucrative trade against pirates and competitors, Portugal's John II sent an expedition in 1481, of which Columbus was a member, to build São Jorge da Mina (Elmina), the first European fort and trading post on the Gulf of Guinea. Thus, when Columbus saw golden objects among the natives of Hispaniola, his hopes rose of founding a new Elmina for gold mines in the island's mountains.[47]

As mariners sailing under the Portuguese flag came to understand the prevailing northeasterly trade winds of the equatorial zone, they developed the *volta do mar*, a technique of sailing northwest into open ocean to ride the winds back home. Far from mainland, they also discovered or rediscovered the Madeira, Azores, and Cape Verde island groups. Like the Italians, who had seized islands in the Aegean and eastern Mediterranean to protect sea lanes, provide themselves safe harbors, and discourage rivals, Portugal claimed the uninhabited Madeira, Azores, and Cape Verde Islands, although Spain wrested away the Canaries.

Sugar plantations with racial slavery

On these Portuguese islands, the system evolved that inspired the plantations of the Western hemisphere based on racial slavery. To cement its claims to the Atlantic islands, Portugal would have to colonize them, a virtually unprecedented project that Portuguese learned by trial and error.

Other colonizing nations would follow the path blazed on Portugal's Atlantic islands. With Italian aid, investment, and advice, the Iberians followed the Mediterranean model closely. Colonizing isolated, unpopulated islands presented more daunting challenges than colonizing long-populated regions with nearby food and supply sources, labor, and trade routes, as Italians had done. On Madeira, King John I applied the feudal model that Iberians had developed to resettle depopulated tracts after the plague and the expulsion of the Moors. He divided the two islands into three hereditary captaincies and gave two on Madeira to two Portuguese and one on Porto Santo to an Italian (Columbus's future father-in-law) on the condition that they people the islands with colonists. Land was distributed according to social rank. Large estates contracted with tenants to settle.

The first settlers arrived in Madeira between 1420 and 1425. Abundant wood, water, and rich soil offered colonists opportunities to advance their situation over what they had had at home. Timber and abundant cereal harvests on Madeira's fertile virgin volcanic soil made for profitable exports to Portugal, which faced shortages of both. But because the island supported few farm animals whose manure could fertilize the soil, initially abundant harvests declined.[48]

Madeira's small farmers and tenants needed a replacement commodity. Genoese merchants brought sugarcane.[49] By about 1450, sugar was so profitable that Madeirans planted it on most of the island's best soils and shipped more sugar to Europe than it had ever seen. Madeira lacked laborers to keep up with demand. The Genoese knew what to do. They and others brought Guanche, Moroccan, Berber, and black African slaves. Cane plantations remained relatively small, more like Cypriot and Sicilian antecedents than the huge American plantation operations of a later era. Forests fell to make way for fields and to boil sugarcane syrup. An extensive system of slave-built irrigation ditches based on Moorish irrigation systems in Iberia brought water to fields and mills.[50] By the early sixteenth century, lack of pastures for herds again led to unfertilized, exhausted fields. Fuelwood likely grew scarce. Sugar production crashed. Madeirans replaced cane fields with vineyards for the fortified wine for which it is still famous, although they could not restore fertility. Slaveowners manumitted or exported their slaves. By 1600, Madeira looked much like any other Portuguese province.[51]

Foreigners dominated the Madeira economy. During sugar's heyday, Genoese and Florentine merchants and bankers directly or indirectly controlled 78 percent of the sugar trade. Genoese carried sugar to Mediterranean markets, while Portuguese, Flemish, and French merchantmen brought sugar to Antwerp for distribution in northern Europe. Genoese settled on Madeira, married local women, and acquired many of the largest estates. Columbus was among them. Columbus first visited Madeira in 1478 as an agent of the Centuriones to export sugar to Genoa. He married the daughter of the captain of Porto Santo, giving him vital connections to the royal court and making him one of the colony's largest landowners.[52]

The profitability of Madeira's sugar industry far surpassed its Mediterranean forebears. The cash-short Portuguese crown took notice and encouraged sugarcane on its other islands, but the Azores were too far north and the Cape Verdes too dry. Spain, too, sought sugar revenue. In the 1490s, Genoese and Madeirans planted sugarcane on and imported slaves to Spain's Canary Islands, where Genoese again came to nearly monopolize sugar production and trade. In the second decade of the sixteenth century, to nurture a sugar industry after the gold ran out, Hispaniola's government sent to the Canaries for experts in mill construction and sought Genoese investment capital.[53]

Around 1480, the Portuguese discovered another island, São Tomé, off the equatorial African coast of present-day Gabon. São Tomé changed everything. The Crown recognized the island's vital strategic value on the route ships took to catch favorable winds back to Portugal from Elmina and close to prime slave coasts, but only colonization could secure possession. Early reports made it sound like another Madeira: a large volcanic island, uninhabited, fertile, well-watered, and heavily forested. Trial plantings produced the largest cane any European had ever seen. Plans were made to settle it under the captaincy system and plant sugarcane to attract settlers and investors.[54]

The Portuguese, however, quickly realized São Tomé was no Madeira and would take extraordinary measures to settle. Few colonists wanted to settle on an island four thousand miles away. Nearly all the first settlers of 1485 died of tropical diseases, which brought the enterprise to a halt. In 1493, the Crown resorted to desperate measures and sent a well-supplied expedition of mainly convicts and children taken from Jewish refugees expelled from Castile in 1492. Survivors of disease acquired immunity, but lack of capital for mills kept sugarcane a minor side product.[55] In 1515, flush with gold from the African coast, the king invested in mills. More capital and trade came from the Jewish colonists, who had connections to converted Portuguese Jews, the so-called New Christians. Jews had thrived for many centuries in the interstices of trade between Christians and Muslims and had established communities around the Mediterranean. New Christians enjoyed commercial links to all major European trading centers.[56]

São Tomé's sugarcane operations soon looked like later colonial America's. Slaves from the nearby mainland manned large sugar plantations, whose wealthy absentee owners resided in Lisbon. Unique to this island, some investors and several wealthy planters were black Kongolese. Kongo had a similar system of slave agriculture, but to grow food crops, not sugar for export.[57] Coastal West Africa lacked soil suitable for sugarcane, or Africans might have themselves gone into sugar production.[58] Owners and investors set up large-scale industrial operations from canefields to mills, run by professional managers and worked by hundreds of slaves to produce a commodity for the international market. For the first time, sugarcane plantations relied on slave labor for both agricultural and industrial phases of production and, also for the first time, all slaves were black Africans. Vastly outnumbered Europeans controlled slaves with a brutality limited only by the ease with which slaves could escape into the forested mountains of the island's interior. However, the tiny European population faced so many destructive, deadly revolts that, by the end of the sixteenth century, São Tomé's sugar boom was over. The island's main business thereafter was the slave trade.[59]

São Tomé demonstrated to Europeans how to colonize. As Barbara Solow has shown, all European powers found it very difficult to attract colonists in America, few of whom wanted to raise commodity crops that would bring wealth to the far-off homeland. Free-labor colonies all failed to make much profit and most failed altogether. Colonizers resorted to slavery to people colonies and make them pay. For this reason Africans would outnumber Europeans in the Americas until the 1840s.[60]

Hispaniola's sugar industry also underwent a boom-to-bust cycle, but under rather different circumstances. Columbus brought sugar to Hispaniola but lacked capital and expertise to set up mills. Colonists were uninterested, anyway. As Hernando Cortés commented on his arrival in 1504, "I came here to get gold, and not to till the soil like a peasant."[61] When gold gave out in the 1510s, authorities started up sugarcane operations, with substantial Genoese participation. Then colonists flocked to Mexico after Cortés conquered the Aztecs. Hispaniola's Taíno peoples died from mistreatment and European diseases. To supply sugar-growers with laborer, slavers raided Carib islands and the Bahamas. New sugar plantations in Mexico and Panama demanded labor as well. Between 1527 and 1548 slavers captured a million or more inhabitants of Nicaragua to man hoes and mills.[62] But death was driving down native populations everywhere. Already in 1520, shipments of slaves began to arrive from Africa. After 1535, Portuguese and Andalusian Genoese traders sent regular slave ships from São Tomé to the Caribbean.[63]

The Spanish Crown, however, never valued sugar much after the conquests of Mexico and Peru. Hispaniola languished as a backwater. Only after mainland silver mines declined centuries later did sugar make a comeback there.[64]

Brazil

Columbus's voyages for Spain created potential conflict with Portugal. To protect their interests and preserve peace, in 1494 the two nations signed the Treaty of Tordesillas, which separated the two empires along the meridian halfway between Portugal's Cape Verde Islands and Spain's Antilles. In 1500, a Portuguese fleet riding the northeast trades before turning east toward India bumped into Brazil on the Portuguese side of the line. Aside from export of a few tropical products, especially brazilwood for its valuable red dye, Brazil had only Neolithic inhabitants with no gold or silver or tradable commodities of interest to Portugal, which neglected its claim for the moment. Then, in 1555, Huguenots (French Calvinists) established a colony at present-day Rio de Janeiro and shocked Portugal into colonizing its claim. The Crown again divided the land into captaincies. Vastly larger than earlier island captaincies, most of them failed and the Crown took them over.

Brazil would be mother of American plantation colonies. After expeditions looking for gold and silver returned with empty hands, colonists in Pernambuco and later Bahia, including New Christians fleeing to Brazil to escape the Inquisition, established an extraordinarily lucrative sugar industry, which both attracted more settlers and paid Lisbon for colonization with profit left over. Land grants turned into large plantations centered around mills, which produced high quality sugar on highly fertile bottomland in the warm, moist climate. Smaller planters and tenants also grew sugarcane and processed it into sugar in the mills of larger neighbors. In the early seventeenth century, mills adopted a far more efficient technology, in which workers passed whole canes back and forth between three vertical rollers, rather than cutting cane into short lengths before crushing. More-efficient crushing created need for more-efficient processing before cane juice spoiled. The single large cauldron was replaced by a system with a battery of cauldrons of decreasing size and increasing heat. Workers ladled increasingly concentrated juice from one to the next. By the seventeenth century, Brazil's sugar production exceeded the value of Portugal's spice trade.[65]

Plantation slavery was very productive, appallingly brutal, and ecologically quite unsustainable. As sugar production expanded, demand for labor rose. For production of commodity crops, slavery proved more efficient and more profitable than free labor ever did. Plantation owners could organize and discipline their laborers, put more of them to work for longer stretches, and produce more sugar (or, later, tobacco, cocoa, coffee, cotton, rice, and indigo). Never had there been so efficient a system for extracting money from the soil.[66]

Such profits came at a monstrous human cost. Indian slaves provided labor until *bandeirantes* (slave raiders) and disease had nearly eradicated native groups far into the interior. By the 1570s, enslaving Indians also had become offensive to Crown and Church. Landowners began to buy African captives brought on crowded, deadly ships from Madeira and São Tomé. After a slow start, thousands began arriving every year. Almost half of all Africans who survived the deadly crossing to America stepped off the boat into Brazilian slavery. Vastly outnumbered, European owners and overseers resorted to violence and terror to maintain control of unfree laborers. Conditions in slave plantations were so grueling and harsh that birthrates trailed deaths in most European plantation colonies.[67]

Sugar's other high price was ecological. Bearing the brunt was the moist, fertile, and ecologically rich Atlantic Forest, which originally extended over two thousand miles along the Brazilian coast and as far inland as rains

Figure 4. William Clark, "Holeing a Cane-Piece," *Ten Views of the Island of Antigua, London,* 1823. Cultivation of sugarcane was a lucrative business that demanded great numbers of regimented laborers, as seen in this lithograph. Along with the group in the foreground, another gang works in the distance, on the rise near the windmill. Workers in the foreground place stakes aligned with a chain to guide the hoeing of square holes. Manure from the penned animals at left will be packed into the holes, where cane will be planted. The windmill for crushing cane shows Dutch technological influence. (Courtesy of the John Carter Brown Library)

could reach. The forest had endured swidden (or slash-and-burn) agriculture for several thousand years, which had reduced its complexity and biomass, particularly after the arrival of the (by most historical accounts) aggressive, cannibalistic Tupi a millennium before the Portuguese. The Portuguese brought a huge number of non-native biota. From Portugal came cattle, citrus, rice, and wheat and grapes for Christian ritual. From Africa came tropical plants like yams, bananas, ginger, and okra. The Tupi liked some crops so well that bananas and sugarcane spread to groups far into the interior who had never seen a European.

Neither colonists nor their enslaved workforce paid much heed to environmental destruction. Tropical forests present the illusion of fertility, but agriculture quickly exhausts tropical soils. Soils along floodplains and rivers proved much more fertile than the rapidly depleted uplands. The Portuguese used native American and African methods to burn clearings, plant in the fertile ashes, and move on when soil gave out. Royal policy encouraged incineration of forest. Since all trees belonged to the crown, landowners could burn them but not harvest them for profit. Rising sugar production demanded fuel for mills that forests increasingly failed to provide—not to mention the need for wood for tens of thousands of boxes to ship sugar abroad. But the forest was huge and diverse and lasted many centuries. In the meantime, colonists were nothing but prodigal with wood.[68]

The mines of Potosí, Zacatecas, and Minas Gerais

Precious metals would exceed sugar as the American colonies' most profitable export. As gold and silver had funded Italian empire and trade, so they subsidized Iberian empires but on a grander scale. American silver and gold accelerated European settlement of the Americas, paid for trade with the East, sped up the global economy with floods of currency, and launched Europe toward industrial capitalism. Lisbon and Seville, however, were no Venice and Genoa, with their profitable local industries and financial institutions. Portuguese and Spanish wealth came almost completely from trade or resource extraction, that is, mining soil, forests, or minerals. Iberian wealth, power, and glory declined along with their empires. Centers of finance and manufacturing grew instead in Antwerp, Amsterdam, and London. There, not in Iberia, the great economic transformations of the Industrial Revolution would take place. Today, Latin America's silver and gold are mostly gone, but local environments will be scarred for many generations to come.

The Elmina of the New World that Columbus hoped for never appeared before his death in 1506. Then, in 1545, the Spanish stumbled on a mountain of silver on a bleak high plain at Potosí, in modern-day Bogota. Two years later, in 1547, they found a hill of silver rising like a "navel" on the barren high plains at Zacatecas, in Mexico.[69] These two places soon far outproduced all the world's other silver mines combined. Nearly two centuries later, Portugal found its own long-sought Elmina, a rich deposit of gold in Brazil. Between 1550 and 1800, the Americas produced 80 percent of the world's silver and 70 percent of its gold.[70]

Potosí silver poured into Spain. As the high-quality silver gave out around 1570, the Spanish introduced the mercury amalgam method of separating out silver from crushed, powdered rock. An open-pit, or opencast, mine in Huancavelica, in modern Peru, provided the necessary mercury. In both Potosí and Huancavelica, labor levies from local populations dug alongside free salaried workers. African slaves were too valuable to use in the mines.

Human lives and the natural world were sacrificed upon a silver altar. Mercury fumes poisoned the air in both Potosí and Huancavelica. Health problems and high mortality were notorious, and Indians slowly left the region to escape service. The soil absorbed so much mercury that residents today still suffer many neurological, developmental, and other health problems from pervasive environmental and atmospheric mercury. By the 1590s, the open-pit mine at Huancavelica could no longer be worked. Workers dug uncoordinated tunnels following the vein of mercury-bearing cinnabar ore. The unventilated labyrinth they created made it difficult to extract ore as well as unhealthy to work in, until in the early seventeenth century experts from Almadén cut ventilation shafts. Eventually, mercury shortages required imports from a new mine in Idrija, shipped via Venice. In the 1630s, after introduction of an improved method of extracting mercury from ore, Huancavelica's production increased and worker health improved.[71] Trees and any other combustible vegetation for miles around both Potosí and Huancavelica vanished into kilns, smelters, and refining works, leaving bare, red, gullied mountains.[72]

The colossal lode at Zacatecas, Mexico, rose from an arid, unpopulated plain. Without local labor to levy, the Spanish enslaved Indians (usually nomadic Chichimec or others taken in "just wars"). African would in time replace Indian slaves and worked alongside Indian free wage-laborers, who came from the south to find work. At Zacatecas the amalgam method was used on a large scale. Large metal bells above roasting amalgam captured most of the mercury, which condensed and dripped down the sides to be reused. Absence of free-flowing streams for water power forced workers to use muscle power from mules or African slaves to provide energy for pumps and refining processes. Miners laboriously carried ore and waste rock on their backs up long tunnels to the surface.

Mining's environmental impact was large and lasting. Hundreds of tons of mercury every year were lost to the air and soil, or washed downstream, to pollute water supplies and riparian areas along riverbanks. Enough mercury, silver, and gold remain in the soil today that companies have been extracting them since the 1920s, releasing more mercury into the air in the process.[73] Workers denuded sparse local hillsides and canyons to get wood for the smelting and amalgam works, cooking, mine machinery, and construction, although the compact quality of the rock required little shoring, sparing the need for mine timbers. Salt for the amalgam process was harvested by labor levies in salt pans to the east. Mercury came from the Spanish mine in Almadén via ship and mule train, but never in adequate amounts, so miners also imported some from Huancavelica.[74]

So much silver entered world trade from these two mines that it remade global economies and helped lay the basis for the modern global capitalism. Spain minted numberless pesos, which served as the standard unit of international trade until the nineteenth century. Spain spent its silver and gold as fast as treasure fleets could bring it. Pesos flew through Europe, throughout the Mediterranean, and into the Ottoman Empire, Persia, and India, usually ending their career melted down into silver bars in China. Like ancient Greek drachmas or medieval Venetian ducats, coinage smoothed and sped trade far from the country that minted them. Pesos encouraged production of commodities from sugar to fine porcelains and bolstered local wealth and capital.

Spain, however, did not benefit in the long run. Even if Spanish America was more commercially minded and less feudal than mother Spain, Spain wanted silver more than any other commodity and neglected other industries.[75] The economy of the Spanish empire remained remarkably self-contained, exporting little besides metals and certain products like the valuable red die cochineal. The herds of sheep that devoured the Mexican countryside produced wool only for domestic consumption.[76] In Spain itself, a feudal mentality devalued trade and industry. Spain's primary product, Merino wool, was raw material for spindles and looms of other nations. Spain completely lacked capitalist structures like companies or firms, banking and credit institutions, or a commercial legal system to enforce contracts.[77]

Genoa was Spain's banker. Spanish silver flowed to Genoa and out again to pay for products from all over Europe and the Near East.

Genoese prosperity depended less on its fleets and more on its bank vaults. But banking chained its destiny to Spanish prosperity. Constant warfare bankrupted Spain nine times between 1557 and 1666. Genoa could not sustain such losses. When its leading merchant houses fell, it had no trading empire to fall back on. Genoa retreated from the center stage of history to its wings.[78]

Then was Portugal's turn. In 1690, a Brazilian slaving party discovered the richest gold deposit in South America, around Ouro Preto ("Black Gold") in the modern state of Minas Gerais ("General Mines"). Soon 450,000 Portuguese immigrants bringing half a million African slaves swarmed in. Gold lay in gravel in streambeds and under topsoil. Africans from the gold regions taught the Portuguese to pan for gold. Standing in cold water under the hot tropical sun was quite unpleasant but healthier than working in silver mines.

Once more, the land paid a high ecological price. After the streams of Minas Gerais were panned out, miners dredged them. Then they dammed and rerouted them to wash away topsoil and separate lighter gravel from heavier gold. Elsewhere, as they had in Africa, slaves dug pits to expose gold-bearing gravel. A once dense forest turned a moonscape, with piles of gravel on a barren, pockmarked land plagued by erosion and floods. Warren Dean notes that the mercury amalgam process was used on lower-quality ores, but how much and with what environmental effects no one knows. Additionally, more Atlantic Forest went up in flames to create fields to raise food for miners. Mule trains brought foodstuffs from an ever-greater distance as agriculture exhausted nearby soil.[79]

Portugal developed its economy no more than Spain. Slavery and coercion marked Portuguese economic activity for centuries. Economic policies benefited merchant and military elites, not the broader population. Gold, sugar, and spices passed through Lisbon on their way elsewhere. No industry and banking developed to generate and spread wealth locally. Portugal's agriculture stagnated. The country remained a nation of peasant farmers until the 1990s.[80]

The Italian empires were built by traders, the Iberian by Crusaders. A plantation capitalism of capital investment in land, machinery, and slave labor did not lead to industrial capitalism. Portuguese sugar-growers never used double-entry bookkeeping or acted like Italian businessmen.

Eighty percent of Brazilian gold ended up in England, Portugal's ally and major trading partner. There, it stimulated trade, banking, and industry, while Portuguese industry starved for capital.[81] In northern Europe, the tale of profit would take an unexpected turn.

THREE

The Wonders of Coal and Machines

Twelve hundred miles north of Lisbon and a thousand north of Genoa lay Glasgow, Scotland, a town with 2,500 people in 1492, on the cool, damp west coast of western Europe's poorest nation. Genoese and Iberian explorers and merchants sailed around the world but rarely or never docked at this port, which exported salted herring.[1] Yet there, in 1765, a young instrument-maker named James Watt discovered how to harness heat energy for tremendous industrial power. No Galileo experimenting at a center of Italian capitalism in Florence or Venice, no Jerónimo de Ayanz tinkering at a hub of trade and empire in Seville or Lisbon, no Christiaan Huygens inventing in Amsterdam or London accomplished this.

A sober, earnest young artisan thinking and experimenting at virtually the limits of European civilization, Watt wrought an energy revolution that would redirect the destiny of the earth itself. In 2000, Dutch Nobel Prize–winning chemist Paul Crutzen and Eugene Stoermer of the University of Michigan first suggested that we have been living in the "Anthropocene" epoch, which he dated to precisely 1784, the year that Watt perfected his steam engine.[2] After 1784, coal combustion increasingly powered the Industrial Revolution and still powers much of the world economy today. However, it also has added about 800 billion metric tons of carbon dioxide to the world environment, which, Crutzen and Stoermer argued, was enough to warm the global climate.[3]

How preposterous, at first glance, that a city like Glasgow should have fostered the crucial discoveries that would bend the arc of capitalism toward its industrial age and toward global environmental crisis. There, on the Firth of Clyde at the end of the Antonine Wall, Roman civilization had had just the faintest frontier presence. While capitalism evolved from roots in ancient Greece and Rome through the medieval Italian republics to the rise of the great Iberian global empires, Glasgow sat on the sidelines of the world economy. Aside from the herring trade, all it could take pride in was a bishop and, after 1451, a university.[4]

It took a remarkable confluence of developments in plantation capitalism, imperialism, trade, mining, manufacturing, and incipient industrial capitalism, as well as a new intellectual and moral force in world history, radical Protestant Christianity, to transform Glasgow into the springboard for Watt's spectacular rise to success, wealth, and fame. In this new, prosperous, expansive Glasgow, Watt befriended Adam Smith, liberal capitalism's great theorist, and Joseph Black, the scientist who discovered carbon dioxide and latent heat. A more fitting trio of attendants at the birth of the anthropogenic global warming can hardly be imagined.

Glasgow anchored one end of a region stretching southward to Manchester and Birmingham which both hosted radical Protestantism and fostered the mechanical innovations that brought industrial capitalism to the world. Out of Britain's western margins away from centers of power, finance, and culture in London, stormy winds of economic change blew, cloudy with coal smoke, and darkened skies around the world.

Moving from the Italian-Iberian center: Dutch capitalism

No Scot would have been in the position to develop a steam engine had the Netherlands and then England not prepared the stage for Scotland's entry into industrialization's spotlight. The Dutch brought the leading drivers of the global economy from Iberia and Italy to Amsterdam. In the Netherlands, they built the first industrial sector that relied on wind energy and fossil fuels and, in the Americas, spread commodity agriculture and plantation slavery. The English founded successful colonies only after the Dutch preceded them and gave them essential assistance. In the eighteenth century, when the door to England's empire opened to poor, weak Scotland, Glasgow finally gained the trade with American plantations that allowed Glaswegians to fund Watt's experiments.

At the time Columbus sailed, the Dutch and English had nearly as far to rise as the Scots. Only about 8,000 people lived in Amsterdam and about 50,000 in London.[5] Dutch capitalism and empire developed first, emerging from a land on which nature had bestowed a challenging mixture of blessings and curses. In the early Middle Ages, in part to escape feudalism, people first began to filter into the cool, damp, breezy, flat, marshy delta where the sluggish Rhine, Meuse, Ijssel, and Scheide snake their way to the North Sea. Ice-Age glaciers, rivers, wind, and sea had left a wide variety of soils and extensive peatlands prone to frequent floods and storm surges. Feudal lords never gained more than a tenuous presence in most of the Netherlands because the grain that supported them did not thrive in the cool, wet climate. Yet those rivers provided easy transportation routes from the ocean to the interior and a long shoreline offered access to lucrative North-Sea herring fisheries and to sea routes to the Baltic, Arctic, and Atlantic.

Keeping water at bay fostered self-government. Farmers dug ditches to drain fields. But peat shrinks as it dries and turns into fens and lakes, so they built low dikes to keep water out. Deforestation in the Rhine and Meuse watersheds worsened the flood peril. Silt settled in watercourses and raised water levels, regularly forcing townspeople to raise dikes. To forestall floods, high tides, and waterlogging, by the year 1000 villagers were banding together to build and maintain dikes, sluices, and canals. Regional water boards coordinated efforts and managed expenses. Canals also expanded the transportation network. Peasant republics emerged in the north, while the counts and bishops who ruled to the center and west encouraged reclamation by comparatively autonomous villagers in exchange for military contributions. Dams to prevent salt intrusion up rivers offered convenient places for seaborne and inland traders to meet. Towns sprang up at the dams, notably Amsterdam on the Amstel River and Rotterdam on the Rotte. By the sixteenth century, nature would hardly have recognized her handiwork. Hardly any portion of the Netherlands remained that might be described as "natural."

Attempts to control nature gave rise to a capitalist market economy and an agricultural revolution. The effort and expense of creating and maintaining farmland could only be repaid if farmers produced high-value goods for the market. Grain imported from east of the Elbe freed those near urban centers to specialize in high-value horticultural and industrial crops and those farther away to make butter and cheese. Varied soil types and easy transportation encouraged agricultural specialization. Farmers experimented with crop rotation and other innovative methods to increase production. They bought manure from dairy farms, nightsoil from cities, and waste from breweries and other industries, and fertilized extensively. These intensive farming practices supported a dense population.

In the constant battle to keep floods off subsiding farmland, the Dutch developed advanced engineering and technology. By 1408, the grain windmill was redesigned to pump water. A century later, well over 100 windmills were draining water from Dutch fields. Expensive and complex, windmills required capital and expertise. Urban merchants invested in windmills to drain lakes and polders to create land to rent or sell to farmers, creating a market in farmland that further promoted capital-intensive, profit-driven agriculture. Wind-powered drainage led to greater investment and improvements in agriculture, while advanced crop rotation and heavy manuring made Dutch fields the most productive in Europe.

The ever-present sea invited the Dutch to fish and trade. Dutch fishing fleets dominated the rich herring fisheries of the North Sea. Flemings in the populous region south of the Rhine imported raw English wool, which a putting-out system turned into fine textiles that they sold to Italian merchants at the fairs in Champagne. The arrival of the first Genoese galleys in Bruges in 1277 inaugurated cheaper maritime trade with Italy. Bruges prospered and might have become the Venice of the north had its harbor not silted up around 1500. Antwerp assumed Bruges's role as northern Europe's financial and trading capital. Portuguese carracks full of sugar and spice and Spanish galleons laden with silver unloaded in Antwerp, not Iberia. By the fifteenth century, the Netherlands was bustling with economic activity and was among the most urbanized regions of the continent.

By the 1600s, Dutch flags flew on over 16,000 ships, more than all other European colors together. With almost half the world's trade flowing through Dutch ports, the Amsterdam harbor filled with a floating forest of the masts of thousands of ships of all European nations. The Dutch innovated important forms and instruments of financing and corporate organization, especially the joint-stock corporation and limited liability. High wages encouraged industries that added significant value, and Dutch products gained a reputation for workmanship and quality. Dutch technical innovation reached a peak, outpacing all other nations in the seventeenth century. High wages attracted poor Germans and Walloons, who built and dredged canals and dug peat. Lack of strong central authority encouraged religious freedom, as the Dutch never had the political unity or power to impose Calvinist uniformity. Uniquely tolerant religious policies attracted talented and skilled immigrants. Immigrant Jewish merchants from the Iberian empires brought capital and trade connections. After 1685, thousands of French Huguenot refugees fleeing the Revocation of the Edict of Nantes came with capital and industrial skills.[6]

Trade abroad fed industry at home, including many energy-intensive industries using fossil fuels. Goods, silver, and profits that came in from the Iberian, then Dutch, and finally English and French empires provided capital to invest in domestic industry. Plentiful, cheap energy encouraged manufacturing. While other nations relied for fuel on exhaustible wood with its high overland transportation costs, heat energy for Dutch industry and households came from local peat or coal floated down the Meuse from Liege or across the North Sea from Newcastle. For the first time in history, a nation relied primarily on fossil fuel (or essentially fossil, since peat renews itself only over long periods). Fossil fuels supported a wide variety of profitable energy-intensive industries: sugar refineries, breweries, brick and tile works, ceramics (famously at Delft), whale oil, glass, distilling, and salt.

Free wind energy complemented cheap fossil fuels. Waterpower is practical only where sufficient water flows and elevation drops (a rarity in a flat country), but windmills could be located where need was greatest. Industrial windmills processed tobacco and made paper for the thriving printing industry. Shipbuilding thrived using lumber from Norwegian, Baltic, and German logs sawn in hundreds of wind-powered mills. Dutch ships cost a third of ships built in England. With cheap energy available for most purposes, the Dutch had no pressing need to develop a greater source of power from steam.

Dutch capitalism made a dramatic impact on the environment, even beyond the extensive water engineering to dry the land and keep it dry. The concentrated sugar refineries of Amsterdam and industry in other towns emitted such quantities of stinking coal smoke that several times authorities cited the "insufferably great sorrow, vexation, and discomfort of the residents" and restricted or prohibited the burning of coal.[7] However, refiners pressured city leaders, who after a time relented.

Industry affected environments far from the Low Countries. Dutch explorers looking for the Northwest and Northeast Passages to the Far East discovered ice, mainly, but also immense pods of whales. Dutch and English whalers soon killed huge, docile bowhead whales by the thousands for their oil for lamp oil, soap, candles, and lubricant.[8] The tremendous quantities of wood needed for Dutch merchant and naval ships, along with pitch and other naval goods, arrived from the forests of Russia, Poland, and Scandinavia. Trees within about twenty miles of rivers were cut, sledded over snow and frozen ground, and floated downstream to be shipped to wind-powered sawmills on the Dutch coast.[9]

The height of the Dutch Golden Age coincided with the depth of the Little Ice Age. Wind patterns favored Dutch fleets in wars with England. Strengthened winds sped Dutch ships to the Spice Islands even as stormier seas raised the risk to sailors. The Little Ice Age caused lands around the world to suffer famine, turmoil, and sometimes political collapse and invasions. Access to easy transportation and reliance on imported food, wood, and fossil fuel allowed the Dutch to prosper and warehoused grain let them profit from others' misfortune.[10]

Political and religious circumstances ignited the explosive expansion of Dutch capitalism, trade, and empire that fractured the Iberian global system and opened the way for other nations to push their way in. Through strategic marriages, the Hapsburgs had acquired the Low Countries by 1482 along with Spain and Austria by 1516. Centralizing policies and heavy taxation alienated their independent-minded subjects. Then the Protestant Reformation arrived, mostly Calvinist, and dropped a match on waiting tinder. From Antwerp to Groningen, rioters invaded churches and destroyed "idolatrous" images. Philip II sent an army to defend order and Catholic orthodoxy, only to provoke a full-fledged revolt in 1566. By 1576, Philip found he had overextended his finances with suppression of the Dutch revolt, wars against the Turks and French, tensions with England, subsidies to Catholics in the French wars of religion, and demands of Spain's extensive empire. When he declared bankruptcy, unpaid mercenaries sacked Antwerp. Merchants, bankers, and Protestants fled to Amsterdam. In the next decades, war sent about 850,000 from Flanders to the Dutch Republic.[11] With them went Flemish commerce and industry and Antwerp's leadership in trade and finance.

The Dutch war continued with periodic truces for eighty years, when Spain finally recognized Dutch independence.[12]

The Dutch got an opening to create an empire of their own when calamity staggered the great Portuguese empire at the pinnacle of its power and glory. In 1578, childless King Sebastian I led the Portuguese army into the disastrous Battle of Alcazar in Morocco and perished with most of the nation's nobility. After a struggle, Philip II assumed the crowns of both Spain and Portugal. The Dutch now treated Portugal as their enemy, too, and pounced. The Portuguese lacked the population, ships, naval power, wealth, and military might to administer and defend a domain spread from Brazil to the Moluccas.

Much like the Genoese, the Dutch privatized their overseas imperial and commercial undertakings. In 1602, they organized the Dutch East India Company (Vereenigde Oostindische Compagnie, or VOC), with a monopoly on the spice trade and power to establish and administer colonies and wage war. The Dutch West India Company (Westindische Companie, or WIC) followed in 1621 as the VOC's counterpart west of Africa. The VOC captured most Portuguese trading forts in south and southeast Asia and by 1641 the WIC conquered much of the northeast sugar-producing coast of Brazil. To secure a supply of slaves for sugar plantations, the WIC seized Elmina, São Tomé, and the Gold Coast and captured the Atlantic slave trade.

Then came the first major Dutch reverses. In 1650, Portugal began a revolt against Spanish rule and re-established its independence. The Portuguese expelled the Dutch from São Tomé in 1648 but never recaptured the Atlantic slave trade or lost Asian colonies. They recaptured Brazil in 1654 and England took New Netherlands a decade later, leaving the WIC in control of only a handful of Caribbean possessions. It profited mainly from the slave trade during its remaining century of life.

After England, France, and other nations closed the doors to their empires, the Dutch economy stagnated. Confined to much smaller domestic and German inland markets, which could by no means absorb Dutch industrial production, Dutch merchants and rulers grew cautious and conservative. Technical expertise drained away as opportunities closed at home and other nations hired Dutch craftsmen and workmen, whose skills benefited foreign economies. Amsterdam, as Genoa had before it, became Europe's financial and banking center. After a disastrous 1780 war with England and the Napoleonic wars, the Dutch economy and empire went into a tailspin. The Industrial Revolution would happen in England, France, Germany, Switzerland, and the U.S., not the Netherlands.[13]

The English follow close on Dutch heels

The British would achieve a more successful, integrated, and prosperous empire than any predecessor and a far higher level of domestic industrial development than the Dutch. England followed Italian and Dutch models of colonization and empire and established coastal settlements around the world to access trade or export precious metals or natural resources, complemented by domestic export-oriented industry. England also privatized empire and trade. Englishmen founded the East India Company with a monopoly in the East in 1600, two years before the VOC but with a tenth of the VOC's capitalization. The English divided their American claims among a host of joint-stock companies, private proprietors, and even colonists themselves. Most would transition to royal colonies.

The Virginia Company established England's first successful colony at Jamestown, Virginia, in 1607, although it nearly failed. The original colonists nearly starved to death. Disease sent most colonists to early graves. Angry Powhatan Indians almost wiped them out in a bloody war in 1622. To the despair of investors, Powhatans had nothing valuable to trade, no valuable plants or trees grew in Virginia, the mountains had no gold, and climate disfavored olives, grapes, or silkworms that thrive in Europe at the same latitude. Virginians turned to commodity agriculture with unfree labor. In 1612, colonist John Rolfe developed a marketable hybrid variety of tobacco. Soon every colonist raised this profitable plant and wanted more land and labor with which to grow it. Colonists acquired or took land from outnumbered and outgunned natives. For labor, colonists imported English indentured servants bound to service for four to seven years, whom they bought, sold, mistreated, and overworked. At the end of indenture, survivors of this mistreatment started tobacco farms of their own. The indenture system proved so profitable and so well suited to colonization that England sought to make every new colony another Virginia. Colonists in neighboring Maryland, founded in 1634, also set up servant-grown tobacco plantations. Colonies on Barbados and other Caribbean islands followed the Virginian model as well.[14]

As fellow Protestant enemies of Spain, the Dutch played a crucial role in the success of early English colonization. Separatists had found refuge from persecution in Leiden for a dozen years before founding Plymouth Colony in 1620. English veterans of the Dutch war against the Spanish organized defenses in Puritan New England. The Dutch shipped, processed, and distributed most English tobacco. They brought the African slaves that increasingly replaced indentured servants after mid-century. Alarmed at Dutch domination of its colonial trade, England imposed a mercantilist system that shut the Dutch out of their empire in the 1650s and 1660s. A series of wars followed. In 1664, the English captured the VIC's colony New Netherlands and renamed it New York. In the East, England took Ceylon from the VOC and captured the India trade, although the spice trade remained Dutch.

A Dutch dramatic triumph over England in 1688 ironically gave its rival a permanent economic advantage. Catholic James II's coronation in 1685 raised the nightmare possibility of an English alliance with Catholic King Louis XIV of France against the Dutch. To prevent this, William III of Orange invaded England in 1688, welcomed by English Protestants. James fled, and Parliament bestowed the English crown on William and his wife Mary, James's sister. Unhappily for the Netherlands, capital followed better interest rates from Amsterdam to London. Dutch investors built roads and canals and upgraded dismal English infrastructure. Dutch and Huguenot immigrants modernized England's banking, finances, and insurance. The Bank of England was launched in 1694.[15] Excellent students, the English would surpass their Dutch teachers.

Dutch immigrants brought with them new agricultural techniques, crops, and dye plants like woad. They boosted the English agricultural revolution, which was already underway since Henry VIII seized and sold monastery lands and created a capitalist land market. The enclosure movement also promoted capitalist agriculture as landlords consolidated smallholdings and plowed former common lands. These developments encouraged greater efficiency and orientation to commercial production.[16] Dutch horticultural techniques made possible London's astonishing growth. A pause in population growth after 1650 eased demand on agriculture while growing cities pushed demand for meat. Farmers turned arable to pasture, where manure from large herds of beef cattle fertilized the soil. England ended its dependence on imports for many commodities and began exporting them to both the Continent and the colonies. Many farmers turned to household manufactures to supplement their income. From them later cotton mills drew their laborers.[17]

Plantation slavery of the northern Europeans: racism, violence, and conservation

The Dutch set the English up in the plantation business. England's opportunity for empire had come when, after 1600, Spain could no longer enforce exclusive claim over the Caribbean Sea. Like Portugal, Spain hardly had the population or resources to rule and hold a vast American empire against determined foes, whom tropical diseases as much as fortifications held at bay.[18] Most islands of the Greater and Lesser Antilles, as well as the American continent north of Florida, cost more to settle and hold than they were worth to Spain. The English, French, Dutch, and even Danes grabbed tempting islands and the Dutch, English, French, and Swedes launched colonial ventures on the continent.

As always, claiming an island was more easily done than making it pay. English colonies tried the Virginia formula of raising tobacco with indentured servants, but tobacco, cotton, and other products proved unprofitable. In 1650, James Drax, a Barbadian planter with Dutch ancestry and connections, visited Recife in Dutch Brazil, studied the complicated arts of raising cane and milling sugar, and set up the first profitable Barbados sugar plantation. When Portuguese Brazilians retook Brazil in 1654, Dutch planters and merchants fled, accompanied by Jews, who rightly feared the end of Dutch religious toleration. Playing the role that the Genoese had assumed in the Iberian empires, the Dutch and Jews offered capital, technical expertise, sugarcane, and slaves to Caribbean colonies of all nations, in hopes of replacing lost Brazilian sugar for idled Amsterdam sugar refineries. When French and English mercantilist policies soon excluded the Dutch from all but illegal and incidental trade, they set up sugarcane, cocoa, and tobacco plantations in their own colonies in Surinam and on a dozen islands.

The Caribbean islands would produce immense amounts of sugar, lay shocking numbers of Africans in their graves, and make colossal profits.

The sugar industry was the main engine driving merchandise around the triangle trade with Europe and Africa. It fed the world's fastest growing economy in the eighteenth century to sate England's bottomless craving for sucrose.[19] By 1775, the English were consuming fourteen pounds of sugar per capita a year, seven or eight times as much as the French or Dutch and about as much in total as the rest of Europe together.[20]

As on Barbados, Virginia and Maryland plantation owners replaced indentured servants with slaves. In 1619, English privateers operating under a Dutch letter of marque had sold Virginians about thirty African slaves captured from the Portuguese. Nonetheless, Virginia did not fully transition to a slave colony until the second half of the century, when the quality and quantity of servants declined and lengthening lifespans in Virginia's unhealthy climate made investment in slaves more attractive. Colonial mainland plantations were never as large or profitable as West Indian operations and would support a far smaller slave trade.[21]

Northern Europeans transformed both slavery, which grew more severe, and plantations, which became more efficient. They lacked the continuous history of slavery from ancient times that Mediterranean peoples had and had never known slave-based commodity agriculture. Influenced by Iberians, whose racial consciousness arose from obsession with bloodlines after Spain and Portugal absorbed populations of Moorish and Jewish converts, the English often regarded sub-Saharan Africans as subhuman and bestial. Caribbean slavery was universally harsh. Like the planters of São Tomé, the more outnumbered Europeans were, the more brutally they sought to intimidate slaves through violence.

The Reformed Protestantism, or Calvinism, of the Dutch, English, and especially the strict Scots made plantations more like factories. After the union of England and Scotland in 1707, hungry Scots by the droves went out into the English empire. Planters often employed them as accountants and overseers. Harsh treatment of slaves followed from the Calvinist attitude toward the will. Because human willfulness led to sin, Calvinists sought to suppress their wills and be instruments of God's will, and in turn expected dependents and servants to be instruments of the master's will, as a machine might be. (Tellingly, in the nineteenth century some anti-Calvinists claimed Calvinists regarded humans as machines.)[22] This notion, alongside Reformed abhorrence of waste and inefficiency, transformed the slave plantation into something like a machine or factory, and, indeed, the efficiency and productiveness of English plantations exceeded those of other nations.[23] One Scot wrote that Scottish overseers were "universally more cruel and morose toward slaves" than others. The wife of a Scottish planter claimed that slaves hated Scots more than other overseers because of their "proverbial" fixation on the work ethic and economy.[24] English and antebellum American plantation managers improved efficiency through detailed bookkeeping and kept productivity statistics for individual slaves. Statistical accounting spread widely in the West Indies.[25] England's capitalist-inflected laws of property also reduced slaves from human to cogs in a machine as much as possible.[26]

People are not automatons, of course, and hardly a decade went by in the sugar colonies without a major slave revolt, often instigated by new arrivals shocked at the amount of labor required of them. As on São Tomé, in Brazil and Jamaica escaped slaves formed maroon communities in interior mountains and forests, from which they raided and attacked plantations.[27] Small islands, however, offered no such havens.

On the other hand, Calvinist and Puritan moralism challenged un-Christian treatment of others and promoted opposition to slavery. In 1596, Dutch traders brought 130 Africans, likely captured from the Portuguese, to the Dutch city Middelburg to sell. The city refused the sale and freed the slaves. Conceived as a vehicle for righteousness, patriotism, and profits, the WIC banned slavery but, after it captured Brazil and faced the necessity of paying off investors, it relaxed this policy and never mentioned it again.[28] New England Puritans and Pennsylvania Quakers also rejected plantation slavery, although they accepted household slaves. Massachusetts Puritan Samuel Sewell wrote the world's first antislavery tract in 1700. The (Quaker) Pennsylvania assembly debated an abolition petition in 1712 and passed a tariff to discourage the slave trade.[29] Scots returning from the Caribbean with enslaved personal servants forced Scotland to confront an issue that had been distant and theoretical. After an initial nonplussed response, by the middle of the eighteenth century, Presbyterian churches condemned slavery and judges ruled that Scottish law forbade it. In England, Scottish judge the Earl of Mansfield decided in the famous Somerset case of 1772 that English common law did not support chattel slavery.[30] In France, by contrast, the principle that anyone in France was free ended in 1716, after which a person's status did not change upon entering the country.[31]

Nature and the plantation-factory

Plantations were precursors of the factories of industrial capitalism, in that capitalists invested in capital assets (processing facilities and enslaved "machines") and natural resources (soil and seed) and used energy (slave and animal muscle, fuel, and sometimes wind, water, or steam power) to produce a commodity for the market. The plantation as factory clashed with human will, as well as another force with a will of its own—nature itself.

Plantation capitalism transformed environments on French Martinique, Guadeloupe, and Saint-Domingue (Haiti), and in Louisiana, in Dutch Surinam, and on English Barbados, Jamaica, and other islands. The French reworked watersheds to irrigate the drier plain of Cul-de-Sac on Saint-Domingue. The Dutch in Surinam built dikes and drained land for plantations and used channels as canals to carry cane from field to mill.[32] Planters introduced non-native tropical plants to feed their enslaved workforce and animals. Bananas, yams, and other plants arrived by way of the slave trade. Guinea grass provided fast-growing and nutritious feed for cattle. Coconuts were probably introduced.[33] Development in the early seventeenth century in Brazil of the three-roller mill and the battery of processing cauldrons, possibly introduced under the Dutch, improved efficiency, increased cane production, and exhausted soil and fuel more quickly.[34] Smaller sugar islands lacked Brazil's vast new lands to clear or huge forests to cut and quickly bumped into environmental limits. Deforestation caused sheet erosion and gullies and sent native plants and birds into extinction.[35] Erosion silted up the harbor of Bridgetown, Barbados, in the 1660s.[36] Island planters faced the expense of importing wood from other islands or the South American mainland.

English planters responded with a series of improvements that made Barbados a hub of conservation innovation. By the 1660s, soil depletion had forced plantation owners to increase their slave force to collect the dung of sheep and draft animals and fertilize fields with it. Small farmers without capital to plant cane made a good living raising animals for their dung to sell to large planters.[37] Planters responded to soil depletion and erosion with the labor-intensive practice of cane-holing (see Figure 4). Slaves dug holes two or three feet square and five or six inches deep and surrounded each with ridges of soil. In the middle they planted

Figure 5. William Clark, "Interior of a Boiling House," *Ten Views in the Island of Antigua, in Which are Represented the Process of Sugar Making*, 1833. Making sugar was a complex, capital-intensive, labor-intensive, and energy-intensive process. To save fuel, a single fire heats a series of copper cauldrons holding cane juice of increasing concentration and quality. Slaves skim scum off the cauldrons into a trough in front, which carries it to the distillery. The slave in the foreground ladles contents of the last cauldron into a pipe that leads to a vat where the sugar dries. Two Europeans inspect the sugar while a man with a scale consults the plantation manager. (Courtesy of the John Carter Brown Library)

cane packed in manure. Cane-holes slowed run-off and checked erosion. As the cane grew, slaves packed more dung around the roots and over time dung filled the holes level with the surface.[38] As forests disappeared from hillsides and fuel grew scarce, Barbadian planters dried crushed cane (*bagasse* or *mill trash*) in the sun and used it as fuel to process sugar and molasses. A few decades later, planters adopted the fuel-efficient "Jamaica train," in which a single fire heated the battery of cauldrons through a flue that ran under each in succession (Figure 5). No consistently effective methods of reforestation developed on colonial islands, particularly since planters found it cheaper to outsource lumber and fuel needs.[39]

For cheap power, Barbadians adopted windmills, probably originally of Dutch design, to take advantage of trade winds. As a seventeenth-century Barbadian slave complained, "The devil was in the Englishman so that he makes everything work; he makes the Negro work, he makes the horse work, the ass work, the wood work, the water work and the wind work."[40] Jamaica, a sugar powerhouse since the English took the island from the Spanish in 1655, had substantial forests for fuel on its inland mountains but no reliable wind. In 1768, a Jamaica planter was an early adopter of Watt's steam engine, burning wood instead of coal, and steam engines soon powered other sugar mills.[41]

Tobacco in Virginia and Maryland also depleted the soil. As in Brazil, extensive uncleared land let colonists put off an ecological reckoning. Large plantations were established on fertile river bottomlands. Smaller planters and farmers tilled poorer, erodible upland soils. Planters large and small girdled and burned off the trees and enjoyed great fertility for about a decade, when time came to repeat the process. Cattle did not need barns in the warm climate and roamed forest commons, their fertilizing manure lost. When yields declined, colonists abandoned fields to recover fertility and cleared fresh tracts. This shifting agriculture was sustainable as long as uncleared forest remained. After the Revolution, upland farmers migrated into Kentucky and Tennessee in search of better land. A crisis of worn-out soil challenged all seaboard slave states after 1800.[42]

In the late seventeenth century, Barbadians with their slaves colonized South Carolina and established rice and indigo plantations. Founded in 1732, neighboring Georgia also raised rice, indigo, and, on frost-free sea islands, cotton. Rice monocultures replaced ecologically rich marshlands. Rice culture's long-term ecological impact was generally light, aside from gradual extermination of the rice-loving Carolina parakeet. Slave ships brought tropical diseases, especially malaria and yellow fever, and the mosquitoes that transmitted them. Disease shortened lifespans all over the Southern colonies and made the Carolina coast a graveyard for Europeans, who lacked Africans' acquired immunities and resistance. The Carolina landscape and climate separated the races, with Africans on coastal rice plantations and Europeans on smallholdings in relatively mosquito-free uplands. Much of the rice fed enslaved workforces on sugar plantations.[43]

Scotland reaches for the fruits of empire

Seventeenth-century Scots could do little more than press their faces to the glass and watch other European powers grab Caribbean islands and plant colonies on the American mainland. Although Scotland and England shared monarchs after 1603, they remained separate nations. English mercantilist policies also shut Scots out of their empire. Wars and political and religious conflict disrupted the Scottish economy. Crop failure and famine stalked the land in the 1690s. The catastrophic Darien colonization scheme of 1698–1700 so bankrupted Scotland that it had little choice but to accept union with England in 1707. Educated Scots engaged in lively discussion about solutions to the nation's poverty. They pushed for advances in science to improve agriculture, in geology to locate valuable minerals, and in political and economic theory to diagnose national ills. Out of this national discussion emerged great works of the Scottish Enlightenment, among them Watt's friend Adam Smith's *Inquiry into the Nature and Causes of the Wealth of Nations*, perhaps the most influential product of this conversation.[44]

With the eagerness of outsiders who had found an unguarded entrance to an exclusive club, after 1707 energetic Scots poured out across England's growing empire. For centuries, Scots by the tens of thousands had emigrated for opportunity and escape from poverty.[45] Now in the eighteenth and nineteenth centuries, Scots disproportionately administered the British Empire, emigrated to the colonies, and dominated shipping and trade. With its superior educational system, Scotland also supplied surgeons, administrators, botanists, foresters, and missionaries.[46]

Caribbean plantations wanted Scots for their labor, talents, and good education. Scottish universities offered better education than Oxford and Cambridge and Edinburgh's medical school had no British peer. The Presbyterian church in addition provided excellent (if unintended) training in accounting.[47] Many Scots who arrived as overseers and bookkeepers for absent landlords went on to own plantations themselves.

Triangular trade and birth of the steam engine

As Glasgow profited from the English empire, the stage was set for Watt's epoch-making machine. The clockwise gyre of currents and winds around the north Atlantic formed a carrousel for commerce, delivering fruits of empire to Glasgow's door. Winds and Gulf Stream currents brought ships from colonial America to Britain's westward-facing ports at Bristol, Liverpool, and Glasgow.[48] To satisfy Europe's growing addiction to nicotine, Chesapeake tobacco plantations joined sugar in driving the triangular trade. There, at the peak of the triangle, while Glasgow's sugar works multiplied, Glaswegians almost completely took over the tobacco trade. By 1750, half of all tobacco entering Britain flowed through Glasgow. By 1772 tobacco made up 80 percent of Glasgow's imports from North America and the West Indies combined. American tobacco planters grew dependent on Scottish "tobacco lords" for both trade and loans, which Scots profitably manipulated, alienating a powerful class of American colonists on the eve of Revolution.[49]

Watt's family prospered from the tobacco trade. Watt was born in 1736 in Greenock, Glasgow's port. His father worked as shipwright and chandler to herring fishermen. He also made navigational instruments, dealt in tobacco and other goods, and constructed Greenock's first crane, which unloaded Virginia tobacco ships. Young James loved tinkering with his father's tools and excelled in mathematics and geometry. After apprenticeship to instrument-makers in Glasgow and London, he returned to Glasgow in 1756. Thanks to a relative on the faculty, the University of Glasgow hired him to clean and repair John Macfarlane's astronomical instruments. Macfarlane, a merchant, had settled in Jamaica and acquired a "considerable fortune" of 5,600 acres and almost 800 slaves. He built himself an astronomical observatory and upon his death in 1755 willed its instruments to the university, from which he earned his M.A. The university hired Watt to repair salt damage from the voyage to Scotland.[50] Watt incurred additional debts to plantation capitalism a decade later, when bankers with interests in the West Indies helped finance his steam engine.[51]

On a fateful day in 1763, the university gave Watt a scale-model Newcomen steam engine to repair. He got it going, but because Newcomen engines do not scale well, it did not run long before stopping. However, even full-sized working Newcomen steam engines consume copious amounts of fuel. Watt determined to discover the source of the inefficiency and to understand the mysteries of heat and steam.[52]

The Newcomen engine had been designed for mines. British miners had raised water that seeped into deep mines with ingenious German wooden machines powered by horses or waterwheels. However, horses could not provide enough power to drain the deep tin mines of Devon and Cornwall, and local streams were few and small. England and Scotland, however, had plentiful supplies of another power source, coal. Taking note of recent experiments with steam and heat, in 1698 Devon military engineer Thomas Savery invented a steam-powered water pump with no moving parts except valves. Although cheap and simple, the Savery engine required so much coal that it was only practical in coal mines, where cheap fuel was abundant. Moreover, since it could only raise water about forty feet, it had to be located far down in the mine itself or, in deeper mines, matched with other Savery engines in series. A heat experiment by Huguenot Denis Pepin inspired Devon ironworker Thomas Newcomen to improve Savery's design. Newcomen wrestled technical challenges for a decade and a half. In 1712, he came up with a design for a more powerful and versatile piston-driven machine that could be sited on the surface.[53] Then, despite many efforts to improve Newcomen's design, steam technology stalled.

Watt consulted Joseph Black, a Glasgow professor who also was studying the mysteries of heat. Black had discovered "latent heat," the large amount of extra heat required to turn ice to water or water to steam. Watt discovered that the Newcomen engine's inefficiency derived from using the same cylinder to heat and cool steam. In 1765, he devised an external condenser for the cooling steam and came up with a engine that was two or three times more efficient. Now, to make his improvement available to benefit society, he needed investors.[54]

Scottish religious culture, engineering, and capitalism

The culture that helped push the Anglo-American Industrial Revolution in directions the Italians and Dutch (or Indians or Chinese) did not go took shape under the enduring influence of Reformed Protestantism in the forms of Puritanism and Presbyterianism. Watt descended from proud Scottish Covenanter Presbyterians, militant allies of the Puritans of England during the English Civil War, in which his great-grandfather was said to have lost his life. His father and grandfather were both kirk elders, a position of responsibility and moral rectitude in Presbyterianism. Greenock, his hometown, carried a reputation as home to strict Calvinism and sober, dour merchants. Watt was active in his Greenock church and all his life identified as Presbyterian. As was not uncommon, his convictions drifted from orthodoxy, but his attitudes and behavior remained moored.[55]

Scotland's strong and aggressive Calvinist Presbyterianism imbued the nation's culture with serious purpose and moralism. As God had not created humans to serve themselves, all had a moral duty to be industrious, productive, useful members of society and to improve their material, mental, and spiritual gifts, as Jesus' Parable of the Talents taught (Matthew 25: 14–30). "Improvement" was constantly on Presbyterian lips—and would be the central word in the first sentence of Adam Smith's *Wealth of Nations.* Waste sinfully abused God's gifts. Calvinists believed that poverty prevented men from using their God-given talents to improve themselves and benefit society. Hence the University of Glasgow was open to the sons of crofters and artisans, and Scotland would be perhaps the first nation to require universal education, as John Knox, the founder of Scottish Presbyterianism, had proposed.

Watt improved his own talents by keeping a "Waste Book" to keep track of expenditures and sources of waste to avoid them in the future. Watt disdained "idleness and mere amusement" and had felt "disinclined" even as a boy "to mix in boisterous play and unmeaning idleness."[56] As a sideline, Watt sometimes repaired and built musical instruments for the public, even as he regarded music as "the source of idleness" and was reluctant to hire people who played them.[57]

The Newcomen engine's waste of coal and steam offended Watt's Presbyterian soul. Calvinist-inspired minimizing of waste and maximizing of economy of operation, improvement, and use of talents not to gratify the self but to benefit the common good (all of them elements of the proverbial Protestant work ethic) drove both the early science of heat and the Industrial Revolution. Reformed Protestant researchers dominated the early science of thermodynamics, among them Black and Watt as well as English Puritan Robert Boyle, Huguenot Denis Pepin, Manchester Congregationalist James Joule, Scottish Presbyterian James Clerk Maxwell, and Irish Presbyterian Lord Kelvin of the University of Glasgow.[58]

Watt was the type of dour Scots engineer that became a nineteenth-century stereotype. A common joke said that if one shouted "Scotty!" into the engine room of any British ship (or the fictional *Enterprise* of *Star Trek*), someone would answer. In 1894, Rudyard Kipling wrote "M'Andrew's Hymn," a homage to the Scottish ship's engineer and one of his best-known poems, which linked Scots, engineering, and Calvinism.

> Lord, Thou hast made this world below the shadow of a dream,
> An', taught by time, I tak' it so—exceptin' always Steam.
> From coupler-flange to spindle-guide I see Thy Hand, O God—
> Predestination in the stride o' yon connectin'-rod.
> John Calvin might ha' forged the same—enorrmous, certain, slow—
> Ay, wrought it in the furnace-flame—my "Institutio."[59]

Glasgow and Ulster, Ireland, marked the western end of a crescent across Europe where Reformed Protestantism held sway: northern Ireland, Scotland, England from southern Lancashire through Birmingham to East Anglia, the Netherlands, the Rhineland region, Alsace, through Switzerland (land of Zwingli and Calvin) and finishing in a Huguenot arc across southern France through Montpellier to La Rochelle. The Industrial Revolution was born and spread along this crescent, where a religious ethos promoted improvements of agricultural methods, financial instruments, and industrial technology. The relatively urbanized core of the Calvinist crescent nursed republics and universities. The mix of Calvinism, cities, self-government, and education fostered economic innovation and activity. Frequent religious wars and oppression sent waves of refugees into a Reformed diaspora in Protestant European and across the Atlantic to Puritan New England, Dutch New York, and a Presbyterian colony-within-colonies in New Jersey, Pennsylvania, and the Southern Appalachians.[60]

A transportation revolution begins

As at many other junctures in history, improved transportation played a central part in the birth of industrial capitalism. A transportation revolution boosted the Industrial Revolution. Manufacturers could never have adopted Watt's steam engine without water access to heavy, bulky coal. Britain enjoyed similar advantageous access to sea lanes as the Netherlands. No British town lay farther than seventy miles from the sea. Although the fall line restricted commerce to short stretches of Scottish rivers, England had numerous small navigable waterways. Thcy did not often flow near coalfields, however.

Inspired by the Dutch, the English began building locks and canals in the seventeenth century. Climate and geography favored canal construction. Plentiful rainfall fell on the relatively flat land that lay under a swath of its most populous regions from York to London to Birmingham to Manchester. In the late eighteenth and early nineteenth centuries, canal building accelerated in response to the desire of manufacturing and urban centers to access iron and coal, deposits of which lay near the surface all over the island. By the middle of the nineteenth century, a dense lacework of about four thousand miles of canals covered England. This superb network of water transportation also linked manufacturers to a significant English domestic market.[61] The combination of water infrastructure and domestic market gave Britain a tremendous economic advantage, in contrast with the Netherlands, with its water infrastructure but small domestic market, and France, with its large internal market but poor transportation network.

The canal boom also reached Scotland and gave Watt some important income as a surveyor. Investors hired him in 1768 to survey and design a canal to bring coal from the Monklands to Glasgow and supervise its construction. In 1773, investors again hired Watt to survey a route for the Caledonian Canal to link the Scottish east coast at Inverness with the west coast at Corpach and save a long, hazardous route around the north of Scotland.

Watt's steam engine and the textiles industry

Both Watt and his steam engine quickly keyed into Britain's growing manufacturing sector. Watt's experiments with steam needed greater capital than surveying could provide. Here Britain's thriving iron industry lent a hand. James Watt's first large investor was John Roebuck of Carron Iron Works in Falkirk. A Dissenter from Sheffield and Birmingham and graduate of the Universities of Edinburgh and Leiden, Roebuck was a restless inventor and entrepreneur. Roebuck earned his fortune with a novel process for manufacturing sulfuric acid, which he sold as a bleaching agent for cloth. No doubt inspired by the iron industry in Sheffield and Birmingham, he set up the Carron works to take advantage of Scotland's unexploited iron deposits. He turned to the coal mines of Bo'ness for coal for coke for his furnaces. When he needed more water to turn his waterwheels and more power than Newcomen engines could provide to drain the mines, he grew interested in Watt's experiments. Beginning about 1767, Roebuck funded Watt's further researches and invested in the patent on his steam engine.[62]

In what ironically turned into a stroke of luck for Watt, Roebuck's Carron iron works could not produce pistons and cylinders with necessary precision for his steam engine. Then in 1774 Roebuck experienced financial difficulties. Matthew Boulton, owner of Birmingham's Soho Manufactory, then the world's largest factory, accepted Roebuck's interest in Watt's patent in payment of debt. Steam engines interested Boulton because Soho's waterwheel faced regular water shortages. Birmingham iron workers were far more skilled than those at Carron. Fortuitously, in 1774 John Wilkinson, another Presbyterian, invented a boring machine for cannons that could make more precise engine cylinders. Watt moved to Birmingham as Boulton's partner. Boulton & Watt sold its first steam engines in 1776, one of which pumped tail-water back to the millpond for Soho's waterwheel, the first factory to use steam-power.

Mines, not factories, were Boulton & Watt's main customers. Boulton foresaw that the market for steam engines in mines would be quickly saturated and urged Watt to develop a model that could deliver rotary power for manufacturing. Watt designed a double-acting steam engine for the constant power and then, in 1784 (Crutzen and Stoermer's date for the Anthropocene), a method to get rotary power from the linear motion of a piston. These and other improvements made the steam engine suitable for factories, whose machinery required rotational power at constant speed to ensure smooth working and maintain product quality. Steam engines let manufacturers conveniently locate factories near labor, transportation, and services, rather than where falling water or blowing wind provided power. They provided energy on demand when droughts or calms immobilized waterwheels or windmills.

Orders for this engine came in fast. By 1800 almost 500 Boulton & Watt steam engines were in place.[63] The firm also sold engines in the

Netherlands, France, the United States, and the West Indies. Watt's adopted city, Birmingham, however, bought only two engines (aside from their own works) before 1800.[64] Despite black clouds of coal smoke from its iron and steel works, the city had little need yet for steam power and was not the cradle of the era of global warming. Steam-powered manufacturing and the factory system took root first in Manchester and southern Lancashire, in the cotton textile industry.

The English cotton industry arose as a consequence of trade with India. In the seventeenth century, European traders imported colorful Indian cloth and sparked a vogue for cotton textiles. The East India Company, following Dutch and Portuguese precedent, set up fortified trading stations and warehouses ("factories") along the Indian coast, notably at Calcutta, Madras, and Bombay. English merchants gave orders to local middlemen, who used the classic putting-out system to get finished cloth from small farmers and household spinners and weavers, although in some cities semi-free weavers worked in large manufactories. The huge and lucrative cotton trade increasingly oriented the Mughal economy toward export.[65] With a relatively tiny number of people in a few coastal enclaves, the English became imperialists almost by chance. Disintegration of the Mughal empire in the eighteenth century created opportunity. Dynastic instability, a gigantic military, and a parasitic ruling class sapped Mughal ability to control, administer, and defend its vast territory. The river of American silver and gold that paid for Indian exports caused inflation and undermined Mughal finances. East India Company officials used diplomacy, bribery, and Indian sepoys to control local rulers and slowly brought the subcontinent under English sway. In addition to cotton fabrics, India under the EIC exported salt, tea, coffee, tropical hardwoods like teak (superb for shipbuilding and more resistant than oak to dry rot and shipworms), silk, and, later, opium and jute.[66]

Europeans found markets for Indian cotton across the world from China to America. Africans exchanged slaves for it. American planters clothed slaves in it. Europeans also discovered the pleasures of this inexpensive, light, durable fabric that took dyes much better than wool or linen and, undyed, was much whiter. Alarmed, wool and linen industries pressed governments in England, France, and elsewhere to ban or impose duties on cotton imports. England allowed production for export only.[67] England had produced wool and linen textiles for centuries but had no knowledge of cotton production. Spinning, weaving, and dying cotton cloth is a complex and laborious process.[68] Two waves of religious refugees established England's cotton industry. In the mid-sixteenth century, about 50,000 Dutch and Flemish Protestants fled to England from the revolt against Spain. Flemish settlers in Manchester and Lancashire, old centers of wool and linen production, imported Mediterranean cotton and wove the first cotton cloth in England. In the seventeenth century, about 80,000 Huguenots immigrated, some due to renewed religious conflict in France after 1620 but most due to the Revocation of the Edict of Nantes in 1685. With German and Walloon Protestant refugees, they brought technical knowledge, capital, and the latest fashions.

Without mechanization, England's cotton textiles could not effectively compete with India's, where the wages of highly skilled weavers were less (and by 1790 much less) than a quarter of English wages and where local raw cotton was very cheap.[69] England's relatively robust patent system encouraged a huge number of small inventions to ease the work of making cotton textiles, which led directly to mechanized factories. Dissenters (that is, Puritans and other Protestant nonconformists with the Church of England) invented most improvements. Lewis Paul, a descendant of Huguenots, developed the first roller spinner and carding machine for the tedious preparation stages. Presbyterian John Kay of Bury, Lancashire, invented several improvements to weaving technology, notably his momentous invention in 1733 of the flying shuttle, which allowed a single weaver to do the work of two men.[70] To help his wife spin, James Hargreaves, a weaver from Blackburn, Lancashire, in 1764 devised the spinning jenny, a hand-powered machine resembling a spinning wheel with eight or more spindles. In 1765, Presbyterian wigmaker Richard Arkwright of Bolton, Lancashire, worked with a clockmaker to contrive a machine that spun stronger thread than the jenny but required more power than a person could supply. After experimenting with horse power, Arkwright drove it with a waterwheel, hence its name *water frame*.[71] With its power requirements, the water frame required substantial capital to set up, albeit little skill to operate. Arkwright built the first mechanized cotton mill at Cromford in Derbyshire on the River Derwent, in 1771, the first of a string of factories that made Arkwright very rich. Samuel Crompton, from an influential Bolton family of Dissenters, combined elements of the jenny and water frame to create the spinning mule in

1779, which made better quality yarn of any fiber, suitable for both warp and weft. The Crompton mule could produce in two hours what it took ten hours to produce by hand.[72]

At first, the Indian cotton industry had nothing to fear from cotton mills. The quality of machine-made textiles lagged well behind Indian hand-made. In the West Indies, the thirteen colonies, the Spanish and Portuguese empires, and Africa, the consistent quality of England's machine-made fabric nevertheless to an extent overcame lower prices of Indian goods. After about 1830, machines achieved such an efficiency that English exports finally undersold Indian textiles and shut down the Indian cotton export industry. By mid-century, English cloth began to take a major share of even the Indian domestic market. English exports also shut down the ancient cotton industry of the Ottoman Empire. Machinery now ruled the world of textiles.[73]

Like the Dutch, the English drew free energy from nature. Waterwheels powered early textile mills. Transition to steam power was slow. Early steam engines could not provide steady power, particularly if the load changed when a machine was added to or disconnected from the power shaft. They were also prone to stopping suddenly. Both problems affected the quality of yarn. Watt in 1782 added a flywheel to smooth out motion and prevent sudden stops, but waterpower remained cheaper and more reliable. Mills were still small in comparison with future establishments and proliferated on the many streams and hills in southern Lancashire, northern Cheshire, and elsewhere in England and Scotland where water descended steeply.

However, waterpower had drawbacks. Drought was a recurring problem. On streams with more than one mill, downstream mills received water at the convenience of those upstream. Finally, owners had to locate their mills where water was sufficient and falling far enough, generally in remote locations far from markets, raw materials, and workers, where whole towns, including living quarters and buildings for service industries, had to be built from available capital. Labor was a problem. Employers needed to attract enough workers to inconvenient locations far from towns. A strike or other labor conflict would leave owners powerless to negotiate or find replacements quickly.[74]

Such issues made steam power attractive, despite its limitations and higher cost. Manchester, with its excellent sea and canal connections and abundant coal supplies, went "Steam Mill Mad," as Boulton reported to Watt in 1781. Boulton and Watt sold their first engine to a cotton mill in 1785.[75] Manchester's rapidly multiplying textile mills were almost entirely powered by steam: two mills in the 1780s grew to 52 in 1802 and 99 in 1830. The same ocean currents that brought American cotton to Manchester carried it to Scotland as well. In 1830, Glasgow had 61 steam-powered textile mills. Outside of the textile industry, steam-powered mills remained rare. Waterpower still had an appeal and its technology continued to advance into the middle of the nineteenth century.[76]

The Industrial Revolution abroad

Industrial capitalism now began its march across the globe. By various stratagems, before the end of the eighteenth century the closely guarded designs of Britain's textile technology found their way to France, Germany, Spain, and many other places where, as in England, textiles had been spun and woven.[77] Swiss, German, and Alsatian Protestants manufactured and smuggled textiles into France during a protectionist ban on cotton. After the ban was lifted in 1759, they set up mills in Normandy. Textile mills also sprang up in French Flanders and Belgium, where coal was abundant.

The first cotton printing firm was founded Mulhouse, Alsace, in 1746. Mulhouse would play a large role in the cotton industry and form the core of one of the most important industrial areas of France. Mulhouse was then an independent Calvinist republic affiliated with Switzerland. Nestled on the Ill among the mountains, Mulhouse benefited from cheap water transportation to the Rhine and Rhone, fields for bleaching cloth, clean water, plentiful wood, lack of tariff barriers, and access to Swiss commercial houses in Basel, Geneva, and Neuchatel. In 1798, the French forced reluctant Mulhouse into France, although its Calvinist factory owners long maintained their local power and independence. Minority status among so many Catholics heightened the Calvinist work ethic of hard work, austerity, and industrial investment.

Mulhouse firms specialized in original designs and colors but needed greater supplies of cloth. The Napoleonic Wars blocked imported cheap British thread and promoted construction of mills using Alsace's plentiful waterpower, although spinning technology remained primitive.

Mulhouse and Manchester learned and borrowed from each other, and by 1812 Mulhouse had adopted steam power. Farther from mines and ports than Normandy, Mulhouse paid high prices for coal and raw cotton and so emphasized quality, not quantity. Its textiles drove the English from the French market and by 1834 Mulhouse was exporting half its production. Its mills modernized and technologically equaled England's by mid-century. French national textile production ranked second only to Britain's by the 1840s.[78]

The Industrial Revolution spread to the United States. Samuel Slater, former apprentice at an Arkwright mill in Derbyshire, teamed up in Pawtucket, Rhode Island, with merchants with capital from the triangular trade and in 1791 opened the first small water-powered spinning mill in the country, followed by other small mills that spun thread and put weaving out to home workers. When the Embargo Act of 1807 and the War of 1812 cut off British textile imports, Slater's mills prospered and proliferated. Slater would also be one of the first American manufacturers to use steam power.[79]

Since Slater lacked access to large amounts of capital or the latest technology, his factories were soon surpassed. Bostonian Francis Cabot Lowell, grandson of a Puritan minister, importer of Chinese silks and tea and Indian cottons and distiller of rum from Caribbean molasses, returned in 1812 from a tour of British mills and constructed an improved power loom. With Boston partners, he built a factory in nearby Waltham that, for the first time, put all operations from raw cotton to finished cloth under one roof.[80] After Lowell's death in 1817, his partners developed a location on the Merrimack River with much greater waterpower potential and easy canal connections to raw materials and markets. They named the new mill town Lowell, Massachusetts, the first of a series of mill towns at falls along the Merrimack.[81]

Industrial capitalism, plantation capitalism, and the environment

The cotton textile industry had far-reaching consequences for the natural environment. Cotton mills operated at the end of a production process that began on American slave plantations, which supplied far more cotton than Europe's first source, the eastern Mediterranean, ever could. Sugar island plantations raised some cotton where more-valuable cane did not grow. Varieties of cotton that grew on the mainland were unprofitable until 1793, when a Yale graduate from a prominent Massachusetts Puritan family, Eli Whitney, invented the cotton gin (or engine) to separate fiber and seeds. Suddenly very profitable, cotton plantations multiplied across the southern U.S. for half a century, often on land from which Cherokees, Creeks, and Choctaws had been removed or driven. American production rose from 1.5 million pounds in 1790 to 331 million pounds in 1830 to 2,275 million pounds in 1860, when the South was growing two-thirds of the world's cotton. About 70 percent went to Britain. British and French mills bought around 90 percent of their cotton from the United States.[82]

Cotton production impacted the environment from field to mill. In the factory that was the cotton plantation, human and animal muscle provided power, and soil the raw material for the product. In stifling heat and humidity, slaves cleared rich bottomlands of thick forests and undergrowth, then plowed, planted, thinned, weeded, wormed, and harvested cotton in an exhausting, yearlong cycle. The warm, wet climate supported the non-native African mosquito vectors for malaria and yellow fever. Poor sanitation and bare feet transmitted parasites like hookworm. Frequent downpours eroded exposed soils, especially on slopes. Since most Southern soils were old, initially high yields declined quickly, except on rich alluvial bottomlands. The initial bounty of virgin soil, even on the richest soil, gave way to declining yields, erosion, and diminished fertility. The expense of conservation and fertilization did not pay as long as unplowed western lands were available. When emancipation in 1865 ended plantation gang labor and led to sharecropping and tenancy, the system collapsed. Sharecroppers and tenants could no longer abandon exhausted land for freshly cleared fertile land and lacked the labor to do so in any case. The South spiraled into poverty and soil into gullied exhaustion.[83] The British, French, Germans, Russians, and Japanese then sought to establish cotton monocultures on the U.S. model in African and Asian colonies, in Brazil, and in Russian central Asia. Soil exhaustion and erosion followed.[84]

The manufacturing end of cotton production produced other environmental impacts. Water-powered mills affected waterways. On the relatively small rivers and streams of England, Scotland, and Alsace, milldams interfered with runs of anadromous fish like salmon, and millponds backed water up into meadows and fields. Dams, weirs, and floodgates interfered with river traffic. Multiple dams on a single stream caused conflict between mills upstream and those downstream.[85] Milltowns needed clean drinking water but dumped human, animal, and industrial wastes in rivers. Downstream communities suffered from epidemics of typhoid fever and cholera. On dye days, dyes stained streams bright colors, although in the era of organic dyestuffs, dyes did not pose major ecological or health problems.

The waterpower potential of the Merrimack River's drainage area was huge, equal to all England's drainage basins suitable for waterpower put together.[86] To ensure adequate and convenient water flow, investors extinguished all others' water and canal rights throughout the basin. The investors also dammed and controlled the lakes at the headwaters, Squam Lake and huge Lake Winnepesaukee. The river flowed when factories needed power and stopped when they closed for the night. It was engineered from source to mouth to run as part of the machinery of cotton manufacturing. The river as public resource disappeared into private corporate interest.[87]

Trade and industry, water and coal, capitalism and the environment

Advances in technology and technique opened the world to capitalism, empire, a rising global economy, intensifying resource use, and greater environmental degradation. The gateway to wealth and power lay now on the Atlantic coast of Europe, whose access to ocean winds and currents opened far more opportunities than in the long age when the Mediterranean was the center of the Western economy. Portugal, Spain, the Netherlands, France, England, Scotland, and even Denmark and Sweden raced and elbowed and fought their way to domination of world trade. The peoples of northwest Europe brought to the imperial enterprise a different experience of merchant capitalism, a Protestant mentality, and the aggressive opportunism of gatecrashers late to the party. The same peoples would fashion an industrial capitalism not quite like any manufacturing that had come before, born in radical Protestant regions and quickly spreading through Europe and North America. Plantations and mills began to alter landscapes around the world.

Scotland had just begun to enjoy the benefits of access to empire when Watt was born. In Glasgow, plantation money, mining needs, the Scottish Enlightenment, and Presbyterian culture came together in a creative mixture. Out of this milieu emerged Watt's efficient, powerful steam engine. James Watt and his partner Matthew Boulton died wealthy men. Mills in over thirty industries—textiles, grains, dyestuff, paper, iron, and others—bought Boulton & Watt's rotative steam engine, with cotton mills making up the bulk of industrial customers. Watt's patent on the separate condenser expired in 1800 but his firm continued to sell steam engines long past his death in 1819 at age 84, lionized as a source of national greatness. The patent's expiration set off an explosion of pent-up engineering improvements on steam engines. Most important was the high-pressure steam engine, which needed no condenser, could be smaller, and was portable enough to power boats and railroad trains. By the 1830s, the Industrial Revolution was in full swing. Capitalism accelerated. So did global environmental impacts. A world far beyond anything Boulton or Watt ever imagined was aborning.

FOUR

Age of Steam and Steel

Prosperity and poverty in a globalized economy

In the town of Dunfermline, across the Firth of Forth from Edinburgh, Andrew Carnegie was born in 1835. His obscure origins gave no hint of his future as the wealthiest industrialist of his time. During a career spanning two-thirds of the nineteenth century, Carnegie would pass through or touch on almost every major aspect of the evolution of industrial capitalism from its infancy in handlooms and local textile mills to its maturity in heavy industry and large corporations. The telegraph, railroad, and iron and steel industries in which Carnegie rose from poverty to immense riches also brought a new era of environmental impact, far greater than earlier industrialization, sometimes called the First Industrial Revolution. Mining for iron, coal, and several newly useful metals increased quickly, as did mining for precious metals, which kept the wheels of global commerce in motion. Industrialization created new demand for gutta-percha and rubber that plantations in former forests filled in expanding European empires in southeast Asia. And coke ovens and steel mills like those of Carnegie Steel blackened skies and lungs and befouled rivers.

William Carnegie, Andrew's father, belonged to Dunfermline's prosperous community of linen weavers, which immigrant Flemish or Dutch weavers had established in seventeenth-century Edinburgh. The development of the Jacquard attachment for looms, which produced appealing patterns, caused production to shift around 1825 from Edinburgh to Dunfermline, for its abundant waterpower for spinning mills. Of exceptional quality and beauty, the town's linen damasks made up an important Scottish export. Fully half went to the United States.[1]

This Golden Age for Dunfermline's weavers ended without warning when the global trade connections that had brought them good fortune suddenly evaporated. Widely separated human and natural events—

Chinese purchases of opium, U.S. politics, a volcanic eruption in Nicaragua—brought the global economy to a halt. People whose income depended on commodity production and trade suffered. Many migrated to escape poverty.

The chain of catastrophe began in the revolutions that broke up Spain's American empire into independent republics early in the century. The new nations did not always maintain the purity of silver in the pesos they minted. China, where most American silver had always ended up, refused the new pesos and paid for Indian opium instead with bills of exchange drawn on the Bank of England. Mexican silver pesos now flowed to the United States, which was also flush with gold that England paid for cotton and investments. In the 1830s, forced removal of Cherokees, Choctaws, and Creeks put vast tracts of rich cotton land on the market. The flood of specie (gold and silver coin) inflated a bubble in land, cotton, and slave prices and drew down specie reserves in major banks. Worse, after 1833, the Federal government deposited receipts from sales of public land in poorly regulated state banks, which loaned more paper notes than their specie reserves could secure. The bubble and the money supply kept inflating.

Meanwhile, the Bank of England suffered a credit crunch. Britain's compensated abolition of slavery in the Empire in 1833 forced Parliament to borrow millions of pounds to pay former slaveowners. Specie reserves of the Bank of England shrank. Alarmed bank directors blamed the United States, raised interest rates, and accepted only specie from houses trading with America. Facing further drains of specie, American banks raised their rates.

Then, in 1835, the Nicaraguan volcano Cosigüina erupted. The ash it ejected into the atmosphere cooled the climate of the northern hemisphere for three years. Both the U.S. and Britain suffered bad harvests.[2] More English gold and silver went abroad to buy grain. To protect dwindling reserves, the Bank of England raised interest rates again to a record six percent. Business investment froze.

To deflate the speculatory bubble, the American government in 1836 issued the Specie Circular. It would accept only specie, not bank notes, in payment for land. Land prices collapsed. Tax revenues fell. States defaulted on their bonds. The global price of cotton dropped precipitously. The large cotton firms of New Orleans went bankrupt. Without reserves to cover their notes, banks closed their doors. The collapsing American economy pulled the global economy down with it.[3]

Beautiful Dunfermline linens had no buyers. To compound weavers' troubles, technological improvements were rendering their skills obsolete. After 1830, factories began installing cast-iron Roberts power looms, far better than prior designs. Hand looms all over the globe fell idle. Weavers struggled on in poverty or left. In 1848, the Carnegies borrowed money from relatives, traveled to Glasgow, and boarded a ship for America. They settled near relatives in Allegheny, Pennsylvania, adjacent to Pittsburgh. The region had been a magnet for Scottish immigrants for a century. Pittsburgh was a burgeoning manufacturing center with excellent water transportation to the inland West and its rapidly expanding economy. Abundant natural resources, still mostly undiscovered and unexploited, lay close by in all directions. It was a place of potential and possibility.

Carnegie and the railroad

A new world of industrial capitalism was rising amid the woods and mountains of western Pennsylvania, but the Carnegies got little benefit at first. William Carnegie disliked or perhaps disdained factory jobs. He set up as a weaver but struggled to sell his linen tablecloths. He must have felt a failure in life. The world had passed him by and left him with useless, obsolete skills. He died in 1855 at age 51.[4]

On the other hand, his elder son, Andrew—twelve years old when he immigrated and nineteen when his father died—took eager advantage of the multitude of opportunities presented by a nation rushing headlong through the early stages of industrialization. As Andrew Carnegie grew up, an Age of Steam and Steel shouldered aside the era of textile mills. In Pittsburgh, which just a century earlier had not appeared on any map, young Andrew Carnegie climbed aboard an economic rocket just as it was taking off. It would lift him higher than any other person of his generation.

Andrew Carnegie enjoyed good fortune on two accounts. First, he immigrated in 1848, on the eve of a global economic expansion. In that year, gold was discovered in California just as the U.S. annexed it, and the century's first great gold rush began. The world emerged from the specie shortage that created the Depression of 1837 and, in fits and starts, roared on into an industrial future. More gold entered the global economy in the next fifty years than in the previous 3,000. There were gold rushes in Nevada and California nearly constantly until the 1880s; forty gold rushes in Colorado, Wyoming, and Montana; 28 rushes in Australia between 1851 and 1894; five in New Zealand after 1857; one each in British Columbia in 1858, Lapland in 1868, South Africa in 1886, and Southern Rhodesia (now Zimbabwe) in 1890; periodically in the Gold Coast between 1877 and 1900; and a mammoth gold rush in the Klondike in Canada and Alaska after 1896.

All this gold changed the global economy. California gold alone contributed $1.4 billion. In 1902, a mining correspondent calculated that gold added about £70 million every year. In 1816, the Bank of England had adopted gold as its single standard. Various factors led many other nations to follow after 1870.[5] The gold standard caused economic problems, however, because it drove silver coinage from circulation and caused deflation. Yet the world economy steadily grew, although at the expense of horrific treatment of native peoples in gold rush regions, tremendous social dislocation and confusion of gold rushes, and environmental destruction from hydraulic mining and mercury.[6]

Carnegie's second bit of luck was to emigrate to a Scottish cultural region, "the most successful plantation of Scottish religious culture in the western hemisphere."[7] Although the Carnegies preferred Swedenborgianism to Presbyterianism, in temperament and values they strayed little from Scottish cultural norms. In and around Pittsburgh lived a host of relatives, Scots, and Scotch-Irish who could lend Carnegie and his family a helping hand.

In Carnegie's first job, he worked twelve-hour days as a bobbin boy at a textile mill owned by a fellow Scot. Young Carnegie was a bright, gregarious lad with only a basic education but a love of reading and a desire for self-improvement. A Scotsman in the company that made the bobbins hired the boy to tend a steam engine. Carnegie was soon promoted to part-time bookkeeper, a skill he had picked up in evening classes. A year later, the Scottish manager of a telegraph company hired Carnegie as a telegram delivery boy. During slack times, he learned to operate a telegraph, which led to a job as a telegraph operator. He did so well that the Scotch-Irish division superintendent of the Pennsylvania Railroad, Thomas Scott, hired him as his personal telegraph operator in

1853.[8] Carnegie left forever the realm of bobbins and textiles and entered a new world of telegraphs and railroads.

The astounding telegraph

In Pittsburgh, a city founded within living memory, Carnegie witnessed the dawning of modern electronic communication and steam transportation. "General Robinson, to whom I delivered many a telegraph message, was the first white child born west of the Ohio River," he recalled. "I saw the first telegraph line stretched from the east into the city; and, at a later date, I also saw the first locomotive, for the Ohio and Pennsylvania Railroad, brought by canal from Philadelphia and unloaded from a scow in Allegheny City."[9]

The telegraph seemed very nearly miraculous as people saw "dead machines . . . move and . . . speak intelligibly, at any distance, with lightning speed."[10] The telegraph's revolutionary impact on the speed of communication stands with Johannes Gutenberg's perfection of printing in 1450 and the invention of writing itself in importance to history and economic development. The first electric telegraph solved a problem for railroads. As railroad traffic increased, coordinating trains going in opposite directions required quick communication between stations to prevent accidents and avoid inconvenient delays for trains that waited for an expected train that was late. With new developments in the science of electricity, inventors learned to transmit codes electronically along wires. In the late 1830s and 1840s, Britain adopted the Cook and Wheatstone two-wire system whereby electronic pulses moved a needle on the receiving end to point to a letter on a dial. In 1837, American Samuel F. B. Morse independently invented a simpler single-wire system using short and long pulses ("dots" and "dashes") as an alphabetic code.

The British regarded the telegraph as a public message system like the mail. The postal service absorbed early companies in 1869. Americans associated telegraphs with private railroads and allowed a single private monopoly, Western Union, to buy all competitors by 1868. Consequently, throughout the nineteenth century, four of every five American telegrams dealt with business, while a similar proportion of British telegrams served personal purposes. Business demand encouraged American companies to string telegraph lines far ahead of railroads, as Carnegie witnessed in

Pittsburgh. By 1851 telegraph lines linked all the nation's major cities from Boston to New Orleans and St. Louis, and in 1861 San Francisco as well.[11] With America's large distances, distinct economic regions, and national market, the telegraph equalized regional price differences, eliminated middlemen, allowed lower inventories, and accessed commodity prices and financial situations. The latest news flashed around the nation.[12]

The telegraph greased the wheels of global commerce as it brought people (and empires) closer together. European imperial powers strung telegraph lines to and throughout their colonies—by 1865, seventeen thousand miles in India alone. Capitals could acquire information from distant colonies and send instructions immediately. Central policymaking grew more effective and local freedom of action more constrained. In warfare, the telegraph contributed to British suppression of the Sepoy Mutiny (or Rebellion) of 1857 and was a crucial factor in the American Civil War of 1861–1865. Business motives, not national interest, spurred successful transatlantic cables after 1866. Underwater telegraph cables stabilized international prices and expanded trade.[13]

While the telegraph performed essential service to capitalist economies, its construction and operation sparked major advances in industrial organization. The telegraph required significant capital, salaried employees, accounting, and tightly coordinated oversight of far-flung operations beyond anything capitalist enterprise had ever required in human history.[14]

The telegraph and the environment: copper

With the telegraph, humanity took the first step toward worldwide adoption of electricity for communication, power, and many industrial processes, with attendant environmental impacts. The "net-work of wire"[15] (as Morse's brother called it) over which Carnegie sent electric pulses when his finger tapped the key required natural resources from around the world.

Telegraphs needed copper wires to carry the electric signals. Thanks to growing Welsh copper production, by 1800 Britain was the world's largest producer and by 1856 supplied half the world's copper. Society paid a high environmental price. Smelter smoke contained poisonous sulfur dioxide, sometimes hydrofluoric acid, and heavy metals like arsenic. It killed nearby vegetation, coated vegetation that animals ate, and bioaccumulated in ecosystems and bodies.[16] In 1862, an observer remarked, "The immediate district of Swansea [Wales], which is subjected to the direct and concentrated influence of the copper smoke, is entirely denuded of vegetation; the hillsides have not a blade of grass upon them, but are converted into a mass of debris of gravel and stones." Wastewater damaged the salmon fishery. After mining ceased, mountains of toxic copper and arsenic slag remained.[17] Copper production declined in Britain later in the nineteenth century and rose in Latin America, especially in Chile (briefly the world's largest producer) and Cuba.

The United States mined its own copper. After discovery of gold in Cherokee lands in the Appalachians led to their forced removal in the 1830s, gold prospectors in 1843 found veins of copper in southeastern Tennessee. The ores were rich enough to repay shipment to smelters in Boston and Swansea. In 1850, a railroad reached Cleveland, Tennessee, thirty-three miles away by plank road. Charcoal from mountain forests fueled a half-dozen smelters feeding copper to Cleveland mills, until the Civil War brought production to an end. With its high sulfur content, Tennessee ore required a roast, in which ore was piled on a cordwood bed and set afire. Each roast produced perhaps a million pounds of sulfur gas. Apart from the damage to vegetation and lungs, wood for roasting and for smelter charcoal denuded surrounding mountains. In the 1870s, wood scarcity ended a postwar era of production.[18]

In Michigan lay the world's largest deposit of native copper, with masses of pure metal as large as fifty by twenty by ten feet.[19] Further discoveries of lower-grade ore and the opening of locks bypassing the falls at Sault Ste. Marie between Lakes Superior and Huron prompted a surge in production into the latter part of the century. After other mines opened in the West, the U.S. produced more copper than any other country. Michigan miners crushed ore-bearing rock at the mine to reduce the volume shipped to smelters. Copious flows of water intensively washed the crushed rock to concentrate the copper and carried waste "stamp sands" downstream or into a lake. Because the ore was relatively pure, smelters did not produce dangerously sulfuric smoke. As smelting technology improved, companies later retrieved the sands, reprocessed them for their copper residue, and dumped the finer sands back in lakes. For every ton of copper taken from Michigan rock, thirty-two tons of sand

and two-thirds of a ton of slag remained at mill sites. Toxic heavy metals in tailings have leached into water and environment ever since.[20]

America's first copper smelters in Baltimore and Boston lay far from the mines of Tennessee and Michigan. A vertically integrated firm combining mines, smelters, and rolling and fabrication mills was founded in Pittsburgh in 1848. Other smelters soon followed in western Pennsylvania, but Detroit, settled by innovation-minded New Englanders and located on copper shipping routes, was smelting 40 percent of American copper by 1861.[21]

The telegraph and the environment: electricity

Until dynamos replaced them in the 1890s, batteries created the electricity that carried dots and dashes across the miles. Batteries worked by bathing zinc, copper, and sometimes mercury in acids or solutions of zinc sulphate and copper sulphate, in cells in teak troughs coated with a marine paint made from rubber and naphtha. Huge numbers of batteries powered telegraph wires. The Central Telegraphic Station of London alone employed 20,000 cells. Acid needed changing twice a week, presumably dumped in the nearest watercourse. Fumes pervaded telegraph offices.[22]

Zinc for batteries (and for industry, after development in 1837 of galvanization) came from mines. Zinc ore usually occurred with lead and sometimes a little silver or copper, so frequently zinc and lead smelters sat near each other. England's zinc came from ancient lead mines along the Pennines, as well as less rich mines in the southwest, Wales, Shropshire, Scotland, and the Isle of Man. A process was developed in the eighteenth century that allowed zinc to be smelted from lead mine spoils. Miners suffered chronic health problems from breathing silicon dust in mineshafts.[23] Smelting produced poisonous sulfur dioxide gas and injurious lead fumes. Zinc mines and smelters in Silesia, Belgium, Sweden, Hungary, France, Spain, and elsewhere in Europe outproduced those in England by mid-century.[24]

Lacking expertise and capital, Americans smelted almost no zinc before 1850, after which production increased from mines in New Jersey and Pennsylvania. In the mid-1850s, Pennsylvania Quakers developed a cheaper way to smelt zinc, using anthracite coal. Production rose.[25] To the west, in the Driftless Area of Illinois–Wisconsin and the Lead Belt of Missouri in the U.S., miners found the world's largest deposit of galena, a source of lead and zinc. After 1799, the first lead smelters appeared in Potosi and Herculaneum, Missouri.[26] After 1820, Southerners with rudimentary mining skills brought slaves to the Driftless Area and dug about 2,000 shallow mines. Henry Rowe Schoolcraft decried the environmental mess: He found himself "winding among pits, heaps of gravel, and spars, and other rubbish constantly accumulating at the mines, where scarcely ground enough has been left undisturbed for the safe passage of the traveller, who is constantly kept in peril by unseen excavations, and falling-in pits."[27] The U.S. exported lead by 1841 and ranked first in global production by 1845, although Driftless production declined because timber for smelters was gone by 1842.

Great quantities of zinc lay deep in Missouri lead mines, which after the Civil War, along with deposits in another lead district in southwest Missouri, overshadowed prewar production.[28] In the latter part of the nineteenth century, other zinc-lead mines opened in the American west, along with the huge Broken Hill deposit in Australia. Mining and smelting of zinc-lead ores led to widespread contamination by arsenic and trace minerals like cadmium, which along with zinc and lead migrates into soils and water and bioaccumulates in animals, fish, and shellfish, which people eat. Miners dumped tailings into streams, which ruined fields, killed livestock and fish, and raised the ire (and lawsuits) of farmers downstream.[29]

The telegraph and the environment: insulation

Copper wires and cables need electrical insulation, which tropical trees in Brazil and in south and southeast Asian colonies provided. In 1847, Werner von Siemens of Germany coated a telegraph cable in gutta-percha, product of a tree native to southeast Asia, instead of tarred cotton or hemp. Europeans had known gutta-percha since 1656 as a hard, rubberlike substance that lacked rubber's bounce or elasticity. In 1832, a Malay worker showed William Montgomerie, a Scottish surgeon for the East India Company, how to immerse gutta-percha in hot water to make it pliable before forming it into useful objects, a process Montgomerie promoted in Europe. British, Dutch, French, and Spanish colonies in southeast Asia and Thailand provided raw gutta-percha that European works made into products that needed to be waterproof or resistant to acid, objects that today would be plastic. After Siemens's discovery, it also coated telegraph wires.

Unlike rubber, which is extracted without killing the tree, gutta-percha resides in heartwood. Groups of workers would brave leeches, insects, tigers, and snakes to reach gutta-percha groves deep in the jungle, build platforms to fell trees fourteen to sixteen feet from the ground, cut incisions into the trunk, and slowly drain the latex, most of which remained inaccessibly in the trunk of the now dead tree. A sixty-foot tree yielded about eleven ounces of latex. By 1900, submarine cables alone had required a total of about 27,000 tons of gutta-percha, for which perhaps 88 million trees fell. Although plantations had begun to appear by the turn of the twentieth century, only widespread adoption of the wireless after 1900 prevented the tree's extermination.[30]

Rubber was another excellent insulator at room temperature, but it softened in warm weather and hardened and cracked in cold. In 1839, Charles Goodyear of Connecticut perfected vulcanization of rubber with sulfur and heat. The process rendered rubber suitable for electrical insulation, as well as for weatherproof fabric, certain medical equipment, belting, hoses, gaskets, and, after Scot John Boyd Dunlop's development of the first practical version in 1888, pneumatic tires.[31]

Demand for rubber rose quickly. Throughout the nineteenth century, wild trees in the Amazon in Brazil supplied all the world's rubber. Tens of thousands of workers in debt peonage braved snakes, malaria, Chagas's disease, and leishmaniasis to tap rubber. Rubber made up 40 percent of Brazil's export revenue at the peak of its production. Britain imported 500 pounds in 1830, 20,000 pounds in 1857, and 130,000 pounds in 1874, when its application to telegraph wiring became common. After 1900, the British and Dutch successfully established rubber plantations worked by coolie labor in southeast Asia, which expanded rubber production dramatically and made it cheaply available. Rubber plantations soon also spread in French Indochina and Liberia. These rubber monocultures had significant environmental impacts. They replaced large tracts of native forests, decreased biodiversity, and caused significant erosion.[32]

Industrialization and imperialism

Southeast Asian colonies thus played a vital role in industrialization. The oldest, the Dutch East Indies, had grown from small beginnings under the VOC until the company's financial problems prompted the Dutch government to take control in 1800. The VOC had only ruled those parts of islands that made profits, but the government continually expanded its zone of control. It also established the infrastructure of development, including all-weather roads, postal service, telegraphs, and railroads, and encouraged the extensive culture of sugar, coffee, tea, opium, cotton, tobacco, rubber, and palm oil, and later developed tin mines and oil fields.[33]

Nineteenth-century Malaysia remained mostly uncolonized, aside from Singapore and a few EIC and VOC trading posts. Thousands of Chinese immigrated, fleeing war and desperate poverty and eager to profit from forest products like gutta-percha. Chinese dominated trade, tin mining, and plantation agriculture but decimated a forest ecology maintained for centuries by light exploitation. England remained content to establish protectorates and intervene occasionally to ensure stability. Late in the nineteenth century, fears of German colonization prompted Britain to extend control and displace the Chinese economically. European industrial demand for tin had risen beyond the capacity of European sources, mainly Cornish mines. The strategic importance of tin mines in British Malaya and in the Dutch East Indies rose. Whether Chinese, British, or Dutch, however, weak governance encouraged quick-profit, irresponsible exploitation of tin, leaving pockmarked and eroded hillsides after ores were exhausted, while tropical forests disappeared up the chimneys of smelters. Open-pit mines, just as ecologically destructive, multiplied in the late nineteenth century. Hydraulic mining silted rivers and frequently flooded downstream cities.[34]

Carnegie, though, would oppose imperialism. He believed it hurt rather than helped business. Carnegie campaigned against American imperialism after the Spanish-American War of 1898. Defeated, he turned his efforts against the British war against the Boers in South Africa just as unsuccessfully. He never ceased to lobby Republican presidents to grant the Philippines independence.[35]

The marvelous railroad

It was a railroad that employed Carnegie as a telegraph operator. The rapid growth of railroads more than the telegraph raised both Carnegie and the world's economy to new heights. Railroads were the terrestrial equivalent of the advance that Portuguese caravels and carracks represented for ocean commerce. Without cheap and reliable ocean and land transportation, industrial capitalism would have starved for resources and markets.

Railroads were the gift of steam power. Steam engines had inspired dreams of steam-driven transportation. Mobile steam engines, however, had to wait for expiration of Watt's patent monopoly and advances in metallurgy. In 1802, Cornish mining engineer Richard Trevithick designed a high-pressure steam engine small enough to power a carriage or boat. Soon locomotives transported coal from mines in Coalbrookdale, south Wales, Newcastle, and Leeds to waterfronts or markets. George Stephenson of Newcastle and his son Robert designed the first locomotives for public railroads. Robert's 1829 Rocket locomotive for the Liverpool and Manchester Railway revolutionized rail transport. The public began to ride its trains the next year, thrilling at head-spinning speeds of thirty miles an hour.

Popular imagination embraced railroads with far more enthusiasm than it ever did canals. Canals still had their place, but railroads were faster, if more expensive, and track could be laid where water was insufficient or grades too steep for canals. Just as Britain's canal-building craze had spread to the Continent and the United States, a similar rage for railroads spread. Belgium had begun its astonishingly quick industrialization in 1817, when the Dutch king and government (which ruled Belgium until 1830) gave John Cockerill, a Lancashire industrialist, capital to construct an integrated ironworks near a coal mine. Cockerill constructed the Continent's first steam locomotive and built its first railroad in 1835.[36] By mid-century, Belgium's government had built the densest railroad network in Europe, which facilitated speedy industrial development in the coal-rich state. Germans soon also designed and built locomotives. Other European nations followed at a slower pace. Outside Europe, Britain began laying track in India. In the second half of the century Europeans built railroads in colonial possessions in Africa and Asia. Other Asian and

Figure 6. Utagawa Kuniteru, Illustration of a Steam Locomotive Running on the Takanawa Railroad in Tokyo, 1873. The railroad transformed economies and daily life around the globe. The completion of the Takanawa Railroad in 1872 marked the beginning of Japan's industrialization. It fascinated the public and was frequently depicted in art. (Metropolitan Museum of Art JP3270. Gift of Lincoln Kirstein, 1959)

Latin American nations by the end of the century were building railroads with European or American engineering assistance and equipment.

Living in a large, lightly populated country with weak or ineffective governments, Americans built railroads at a fever pitch without central planning or regulation, often unsafe and sometimes unneeded. In 1825, companies began building tramways powered by horses or English locomotives to transport coal or quarried stone. Passenger railroads appeared four years later. Soon Americans produced their own locomotives, which in this heavily forested nation generally burned wood instead of coal. By 1840, the United States possessed more miles of track (3,328) than all Europe combined (1,818), a gap that only widened, with over thirty thousand miles of track by 1860 in the U.S. alone. Philadelphia, which lacked water routes to the growing west, emerged as a leading hub, especially after the Pennsylvania Railroad linked Philadelphia over the mountains with Pittsburgh in 1854.[37]

The steam engine also began to replace the ancient technology of wind-driven shipping. Steamboats could sail upstream, in a calm, or against an unfavorable wind. In 1807, Robert Fulton designed and built the first successful commercial steamboat, powered by a Boulton and Watt engine, for the 150-mile trip up the Hudson River from New York City to Albany, terminus of the Erie Canal. America's unequaled network of navigable rivers transformed into highways of commerce. By the 1850s, over 700 large steamboats served the immense watersheds of the Mississippi, Missouri, and Ohio Rivers.[38]

A few decades after Fulton, steamships began plying the seas. No longer reliant on ocean currents and winds, they sailed far shorter, quicker routes. Their adoption was slow because they had to carry their own heavy, bulky energy source and refuel at the end of every voyage. These drawbacks encouraged development of more-efficient engines and use of higher-energy anthracite, a type of coal of almost pure carbon that burns hot and slow, leaves little ash, and is almost smokeless. Anthracite deposits, however, rarely occur convenient to shipping lanes. Britain and other colonial powers had to ship coal to strategically placed coaling stations. Until late in the century, most ocean-going steamships carried masts and sails for auxiliary power to save fuel and ensure against coal shortages.[39]

The impact of steam-powered transportation on the global economy was monumental. In the U.S., trains and steamboats reorganized

Figure 7. Currier & Ives, "The Progress of the Century," 1876. Celebrating a century of change since American independence, this lithograph illustrates how dramatically industrialization had transformed everyday life. The press powered by a leather belt and pulley turned by a steam engine prints books and newspapers cheaply, expanding literacy. The telegraph takes center stage, while an open doorway frames a view of a steam locomotive, steamboat, and steam-powered mill, all emblems of a new steam-powered era of civilization. On the telegraph tape, quotations from the Bible and Daniel Webster extol Christianity and the preservation of the United States in the Civil War. (Metropolitan Museum of Art 63.550.377. Bequest of Adele S. Colgate, 1962)

the economy. Various regions of the country specialized economically. Southern slave states produced cotton for New England's and Britain's mills. Grain from Midwestern states fed the South and the growing cities of the East. Much American grain and most of its cotton also went to Great Britain. Railroads and steamboats tied almost every American into the market economy. Merchants who once had looked out across the Atlantic for markets more and more turned their gaze inland.[40]

Steam and steel

The Pennsylvania Railroad would be Carnegie's path to the steel industry. Nothing promoted expansion of and innovation in the iron and steel industry like railroads. Heavy iron locomotives pulled cars with iron frames and steel tires and rode on tens of thousands of miles of steel track and over iron bridges and viaducts. But railroad demand was only the first in a rapidly expanding list of new and growing industrial needs. Steam engines went to work as steam shovels, steam rollers, steam tractors, and steam-powered factory tools. After the 1830s, shipbuilders experimented with iron hulls and ship components and at century's end had abandoned wood for most ships. Cast-iron implements, tools, and machinery proliferated in daily life in Europe and America. Domestic uses for iron and steel products also multiplied and transformed life and work. Iron bedframes and bedsprings deterred insects. Iron stoves, tools, and gadgets invaded kitchens. Iron farm machinery revolutionized agriculture and decreased dependence on laborers. Steam engines ran such farm equipment as threshers. By the end of the century, steam tractors crawled across American fields. Farm productivity rose rapidly. To meet the demand, iron and steel industrial production had to grow. A new Iron Age was born.

Always ready to seize opportunities, Carnegie supplemented his duties as Scott's personal telegraph operator by acting as his secretary and assistant as well. Along with the job came chances to invest in some behind-the-scenes deal-making between the Pennsylvania, its construction firms, sleeping car manufacturers, and other companies with which it did business. As Carnegie learned the business and took on greater responsibility, he was promoted to Scott's position as superintendent of the Pittsburgh division when Scott was promoted to vice-president. Carnegie was twenty-four.[41]

Carnegie could have had no better business education than the office of a railroad. Railroads were the world's first big businesses, larger than telegraph companies, and later large businesses borrowed their methods. They dominated mature industrial capitalism, the Second Industrial Revolution. Every railroad company faced a complexity of organization, control, and financing beyond that of any previous endeavor in human history. Companies had to raise capital for track, right of way, rolling stock, terminals, ticket offices, freight facilities, railyards, and roundhouses; insure everything; hire and supervise engineers, firemen, brakemen, conductors, porters, agents, clerks, watchmen, switchmen, laborers, repairmen, and telegraph operators, few of whom in the early days had ever worked regular hours for wages; hire salaried mid-level and senior managers, lawyers, and accountants long before any professional schools trained them; grade roadways, lay tracks, build bridges, and bore tunnels upon which lives depended, and keep them in repair and when necessary replace them; buy locomotives, freight cars, and passenger cars; keep trains fully booked, fueled, lubricated, and in repair; have trains available when needed and on schedule; prevent trains from colliding or interfering with one another; and advertise and solicit passengers and freight customers. Innovations in accounting and finance in the 1850s and 1860s became standard for all railroads by the 1880s. Within a few decades, positions professionalized along industry-wide standards. Organizational improvements promoted more efficient railroads and made possible larger and heavier engines and cars, better rails, and constantly improving technology. More trains moved faster and more safely. The benefits of cooperation convinced American companies after 1861 to standardize equipment and gauges so that passengers and freight could move most anywhere by rail without laborious, expensive transfers. Here they followed Britain's lead, whose Gauge Act of 1846 standardized rail gauges.[42]

The Pennsylvania ranked among America's best-run railroads. The company used detailed cost accounting, invested in capital improvements, and kept its lines in good repair. The business education that Carnegie got there served him well during the American Civil War. The Pennsylvania's strategic position as the carrier to the Virginia front kept its lines busy and profitable. Carnegie handled admirably the difficult management of flows of soldiers, resources for industry, and war material. The experience further prepared him to go out on his own.[43]

The new Iron Age

In 1862, Carnegie took his first steps out of railroads and into the iron and steel business when he, three engineers, and two former bosses organized the Keystone Bridge Company to replace inadequate or aging wooden bridges with wrought-iron ones. In 1864, he invested in the new Union

Iron Works to build a rolling mill to supply iron to Keystone. A year later, he was president of the firm. At war's end, Carnegie was a rich man. He resigned from the Pennsylvania and set off on a year's tour of Europe.[44]

In Europe, Carnegie made side trips to inspect ironworks, including a stop in Birmingham. Birmingham's iron and steel industry had been a hub of industrial innovation since the eighteenth century and was the place to observe the latest technology and techniques. Close to iron and coal mines in the Black Country, the city enjoyed access to abundant charcoal; clean water; transportation to Manchester, Bristol, and London; and labor from nearby populous counties. A former Puritan stronghold, Birmingham's lack of a local lord or city charter attracted Dissenters after the Restoration, kept it free of guilds and fostered an industrious, bustling, innovative economy.[45] Quakers, who owned perhaps half England's ironworks, made many of the innovations.[46] Quakers Abraham Darby and his son in Coalbrookdale west of Birmingham between 1705 and 1750 perfected the use of coke to make quality iron. The practice spread as charcoal prices rose in the eighteenth century.[47] By many other gradual improvements and discoveries, the quality and quantity of English iron rose. By the mid-nineteenth century Britain was producing more iron than any other nation.[48]

After Abraham Darby III, in consultation with John Wilkinson, built a pioneering iron bridge near Coalbrookdale in 1781, iron bridges proliferated for canal aqueducts and then railroad bridges. In 1797, a textile mill was constructed with iron columns and beams, which allowed more floor space and larger windows. The 1830s witnessed the arrival of cast-iron building decorations and railings, and in the 1850s cast-iron building facades grew popular.

Americans were slow to adopt British innovations. Iron producers used traditional methods and relied on plentiful charcoal instead of coke. Demand for iron in a country with so much wood remained low. Early American factory machinery was made of wood, not iron. New England textile mills imported most of what iron they did need and rarely owned steam engines.[49] A rapid transition of the iron industry began around 1840. In the Lehigh Valley north of Philadelphia lay the world's largest concentration of anthracite. After 1825, a canal network and then railroads connected the mines with markets in Maryland, New York, and other east-coast markets, which had no other ready source of coal.[50] Iron manufacturing spread rapidly in eastern Pennsylvania and farther northeast. The number of American stationary steam engines surged from around 100 in 1832 to over 900 six years later. New England mills now added steam engines for greater and more reliable power.

Coal and steam promoted the rapid spread of large industrial enterprises. In 1830, most large American factories made textiles. Twenty-five years later, coal-fired factories from Maine to Maryland were making a wide variety of tools, implements, utensils, farm machines, and many other items. In Pittsburgh, abundant coal made steam engines so cost-effective that, when Carnegie was young, steam powered Pittsburgh's textile mills, ironworks, a glassworks, and a steam engine factory. By the 1850s, writes business historian Alfred D. Chandler, "the output and technology of American factory production was indeed so impressive that the British government began to send experts to study American industrial techniques."[51]

Under Pittsburgh ran a huge seam of bituminous coal, a softer, smokier coal than anthracite. At the time the Carnegies arrived in the 1840s, Pittsburgh's diverse industries were increasingly concentrating on iron. Beehive ovens to bake coal to high-carbon coke proliferated and filled the valleys of the Allegheny Plateau with smoke. Most of the coke came to Pittsburgh furnaces by the region's numerous rivers or by rail. An English visitor in 1867 famously described "the most striking spectacle we ever beheld" from atop a hill overlooking Pittsburgh:

> The entire space lying between the hills was filled with the blackest smoke, from out of which the hidden chimneys sent forth tongues of flame, while from the depths of the abyss came up the noise of hundreds of steam-hammers. There would be moments when no flames were visible; but soon the wind would force the smoky curtains aside, and the whole black expanse would be dimly lighted with dull wreaths of fire . . . If anyone would enjoy a [striking] spectacle, . . . simply [walk] up a long hill to Cliff Street in Pittsburgh, and [look] over into—hell with the lid taken off.[52]

Discoveries of huge iron deposits in Ohio, the Iron Mountains of Missouri, and one of the world's richest iron deposits in formations around Lake Superior sent iron ore by boat and rail to mills on western Pennsylvania's coal seams.[53]

Americans start making steel

Carnegie played an important role in the rise of the American steel industry. After his European jaunt, he returned to his investments in telegraph companies, the Pullman sleeping car company, and Keystone Bridge, which was then providing iron for Eads Bridge in St. Louis, the first to cross the Mississippi. After 1870, he traveled frequently to London and New York to sell bridge and railroad bonds. Europeans looking for good investments purchased great numbers of American railroad securities. In New York City, the funding needs of railroads gave birth to investment banks, which perfected all the modern instruments of finance, including marketing and speculation. The New York Stock Exchange took its modern form. A large, sophisticated capital market had grown in New York, which after the Civil War could fund both railroads and industry.[54] Carnegie was in the thick of it and building up his fortune from one insider deal after another.[55]

Returning to Pittsburgh in 1872, Carnegie slowly sold off all his investments except iron. An 1870 tariff on imported steel made American steel an attractive investment. Carnegie's iron business expanded and grew more profitable. Carnegie's greatest talent as a businessman was his ability to attract and keep excellent managers and engineers. He himself knew little about making iron, although all his trips to Europe since the middle 1860s included visits to iron and steel works and discussions with owners.[56]

One of those tours, to Henry Bessemer's steel mill in Sheffield in 1872, impressed Carnegie mightily. At the base of the Pennines seventy miles north of Birmingham, Sheffield had ranked as the world's leading steelmaker since the end of the eighteenth century. The city had long made steel cutlery with high-quality imported Swedish iron. Bessemer, a descendant of Huguenots, devised a process in 1855 that involved blowing air through molten pig iron in a huge narrow-mouthed vat, or converter, to burn off carbon. Bessemer steel became the best in Britain. Railroads soon realized that his steel made superior rails. Bessemer's steel rails cost twice as much as iron but were far more durable.[57]

American demand for steel rails shot up after the Civil War to replace iron rails worn out under heavy wartime traffic and to build transcontinental railroads across the West to the Pacific Coast. To meet the demand,

British steelworks multiplied from Wales to Lancashire to Glasgow. The American need for rails was so great that steel imports remained high even after the tariff of 1870. American ironworks experimented with the Bessemer process. Connecticut-born Alexander Lyman Holley adapted the Bessemer process for American conditions, making it cheaper and far more efficient, although sacrificing quality for quantity. Holley designed eleven of the first thirteen Bessemer plants in the United States.[58]

American railroads bought Sheffield steel railroad tires until a German competitor in the steel business appeared, Alfred Krupp, a Reformed Protestant in Essen, in the Ruhr district of Prussia.[59] Krupp's father, Friedrich, a small ironmaker, had taken advantage of absence of British competition during the Napoleonic Wars to develop crucible steel. Alfred perfected the process and later adopted many English innovations, erecting the first Bessemer works on the Continent. By the late 1870s he was producing rails, railroad axles, propeller screws for ships, gear shafts, locomotive parts, and steel artillery. Krupp invented a method to produce seamless railroad tires and sold huge numbers in the rapidly expanding American market. He moved to buy up iron mines across Europe and to vertically integrate. Innovative and globally competitive, by the 1880s Krupp was the largest private company in Europe.[60]

Railroads push back industrial capitalism's frontiers

Not only by stimulating iron and steel industry did railroads advance the progress of industrial capitalism. Most visibly in the United States and Russia, the webs of terrestrial transportation routes they laid expanded industrial capitalism's reach and accelerated exploitation of remote resources. In the U.S., the treeless, dry, almost riverless lands west of the Mississippi defied European settlement or development. Railroad lines snaked from Chicago out onto the plains. Capitalism and environmental disruption followed, although in Russia, weaknesses in the railway system limited its economic and environmental impact.

The railroad nearly drove the American buffalo (bison) into extinction. Factories used leather belts to transmit power from overhead shafts to machinery on the floor. Cattle in the U.S., Latin America, South Africa, India, and South America supplied most of the leather. Tannins for tanning came from oaks and hemlocks from around the world, which led to the destruction of American hemlock forests in mid-century.[61] Around 1871, someone in London or Germany devised a process to tan bison hides. A market arose for the skin of an animal that required no investment to raise, just as the depression of 1873 released an army of unemployed looking for ways to make money. In the next decade, hunters killed almost all of the perhaps ten million buffalo still roaming the Great Plains and shipped the hides east by rail for tanning or export.[62]

Railroads turned the Great Plains grasslands first to cow pasture and then to wheatfields. Texas cattlemen needed a way to get their herds to market. The opening of railheads at Abilene, Kansas, and other towns attracted the great cattle drives of the 1870s. Rail cars brought the animals to Chicago, where large packing companies used "disassembly lines" (models for Henry Ford's assembly line) to butcher cattle quickly and efficiently. Newly designed refrigerator cars carried cheaply produced beef to destinations across the country.[63]

Without government subsidies, transcontinental railroads could not cross hundreds of miles of plains and mountains where no paying customers lived. The Federal government gave thousands of acres of public land to most transcontinentals, which they sold to farmers. Farmers who bought the parcels produced immense quantities of grain and needed railroads to get it to market. Railroads shipped the grain to the grain elevators of Chicago. There, large milling companies ground it to flour and shipped it by rail to consumers across the country.[64]

In addition to butchering cattle and milling grain, Chicago fostered manufacturing. Railroads wanted freight to fill empty cars on their outward trips to the plains. With no local labor force on the sparsely settled plains, farmers eagerly bought reapers and other machinery from Cyrus McCormick's Chicago factory. Deforestation of Wisconsin fed Chicago's lumber mills, which sent lumber for homes and barns to farmers on treeless prairies. Montgomery Ward and Sears, Roebuck sent farmers catalogs filled with consumer goods, all manufactured or warehoused in Chicago and shipped by railroad. To service all of this, a vast banking sector grew up in the city. Immigrants thronged into a city with so many opportunities. Chicago's official population exploded from 4,470 in 1840 to 2,185,283 in 1910.[65]

Russia, too, wanted railroads to open resources on its immense and remote Siberian frontier. Moscow dreamed of becoming the East–West trade emporium for "the silk, tea, and fur trade for Europe, and the manufacturing and other trade for the Far East."[66] The government built railroads deep into Siberia and, thanks to French and British investors, all the way to the Pacific. The Trans-Siberian Railway connected resources to growing industrial centers. It crossed major watersheds, connecting Siberia's rivers with a national transportation network. It encouraged the partly spontaneous settlement of European Russians escaping poverty and limited opportunities. It attracted scores of geological expeditions to map Siberia's mineral wealth. Its construction stimulated the Russian iron industry, particularly in the Urals, although eastern sections relied on American steel. In the long run, however, poor design, mismanagement, political and bureaucratic hurdles, and inadequate capacity thwarted Russian ambitions for Siberian development. Mining and industry failed to develop in Siberia. Environmentally, the most obvious impact was deforestation for railroad fuel and construction and for needs of new settlers, along with more frequent forest fires ignited by sparks from train engines.[67]

Carnegie steel

Carnegie built his giant steel business from his relations with railroads. In 1872 Carnegie organized the Edgar Thomson Steel Works, named with no subtlety for the president of the Pennsylvania. He broke ground the next year on Holley's design for a new, state-of-the-art steel plant situated between river transportation on the Monongahela River and rail transport on the B&O Railroad, so that Carnegie could negotiate the best freight rates and preserve his independence from the Pennsylvania.[68] Holley's efficient designs and innovations allowed the plant to produce more and cheaper rails than anyone else. The layout made movement of raw materials into the plant as easy as possible and placed buildings according to the needs of production. Holley designed a way to capture the white-hot heat from the Bessemer converter to reduce coal consumption. By design, the plant had to produce steel continuously.[69]

In 1873, Jay Cooke went bankrupt building his transcontinental line, the Northern Pacific. The American banking system went down with him, and the effects spread across Europe. By selling off his remaining non-steel investments, often at a loss, and convincing bankers he and his partners were good businessmen and not speculators, Carnegie kept building his new steel plant. The Thomson Works began producing rails in 1875. As his plant was designed to continuously produce steel, he accepted low prices as the depression deepened. He paid his workers well to get good work from them and to keep unions out. The disastrous labor violence of 1877 did not affect Carnegie's plant.[70]

Carnegie charged ahead of the competition. He was always first to mechanize and innovate. Carnegie understood that completion of the transcontinental railroads would lead to declining demand for rails. In the early 1880s an economic slowdown depressed railroad purchases. Carnegie bought the distressed nearby Homestead Works in 1883 and retooled its production from rails to structural steel, installing open-hearth furnaces, which produced better steel more reliably than the Bessemer process.

Carnegie weathered the depression of 1893 and the decline of rail orders better than most other steel companies and emerged the nation's largest producer of structural steel. The solutions that American cities adopted to address problems of rapid growth all required steel: elevated railways, subways, and that American invention, the skyscraper. The skyscraper's steel frame permitted thin, light walls with lots of windows. In 1890, Carnegie added the Duquesne plant, the most modern in the world. By the end of the century, Carnegie's four mills produced more than half as much steel as Great Britain and more than the combined production of Germany, Belgium, and France.[71]

Industrial capitalism's global offspring

Industrial capitalism took various forms as it spread around the globe. Only in Great Britain and the United States did industrial capitalism grow almost entirely from private initiative and investment, with government encouragement but without direct government investment or oversight. Everywhere else, the state played a more active role.

In France, smokestack industrialization did not take off, despite efforts by government and others to build an industrial base by stealing British technology and hiring British technicians. France lacked large, easily accessed coal deposits. Shielded from international agricultural competition by high tariffs, peasants held onto smallholdings they won during the Revolution and did not mechanize or consolidate. Capitalist agriculture did not develop. Peasants remained on the land and did not migrate to cities to provide a labor pool. Employing German Catholic workers, Mulhouse turned into a little engine of French industrialization. When the depression of 1837 crashed textile markets, the city's leading Calvinist families diversified and invested in manufacturing of sewing thread, wallpaper, textile machines, steam engines, turbines, railroads, locomotives, dyes and chemicals, and later automobiles, electric power, banking, petroleum, and plastics. To take advantage of its plentiful waterpower, they established factories on the west slope of the Vosges in Belfort, which became an industrial center when Germany annexed Alsace after the Franco-Prussian War of 1871.[72]

German industrialization had its origins in the sixteenth century in the Rhineland, where Flemish and Dutch religious refugees with trade connections and access to Dutch capital established a textile industry. Uncompetitive in the mass market due to transportation costs, they produced high-quality woolens and silks for German and Russian markets. Industry mechanized and led Germany's economic sectors through the first half of the nineteenth century. Rhineland industrialization prepared the way for heavy industry in the coal-rich Rhine–Ruhr region. A German customs union in the 1830s, railroad construction beginning in the 1850s funded by private capital but directed by government officials, and German unification in 1871 fostered industrialization. Ruhr coal was well suited for coke, but iron was imported. A Protestant commercial and industrial elite employed a German and Polish Catholic workforce. A powerful German military enormously stimulated iron and steel manufacturing. Flush with military funding, Krupp became a major industrial innovator. Germany straightened and channelized the Rhine River, making it little more than a huge canal to serve the nation's trade and industry, especially to bring barges of iron to the mills. This changed the river flow and forced Switzerland upstream and the Netherlands downstream to replumb the Rhine as well.[73]

Some nations industrialized to avoid becoming colonies or falling into economic dependence on western Europe and America. Humiliated by defeat in the Crimean War, vast, profoundly rural Russia set out to modernize. Aware that its economy depended on exporting grain, lumber, and iron to the West to import luxuries for aristocrats, the Russian government created banks, hired engineers from the West, and raised tariffs to build a railroad network and the coal and iron and steel industries to serve it. French and British investors looking for high returns supplied most development capital, as they also were doing in the United States and in their colonies. The tradition of government-sponsored industrialization continued under Communist rule. Soviet leaders pushed heavy industry in the first Five-Year Plan of 1928 out of fear of invasion from the industrialized capitalist nations. Shut off from the world economy, Soviets paid for industrialization by squeezing the countryside to sell grain abroad (and virtually anything vendible) to buy equipment and technology. The Great Depression so complicated the plan that they repressed the *kulaks*, collectivized reluctant peasants, and ignored drought-induced famines. The coal-rich Donets region of the Ukraine, where Welsh industrialist John Hughes founded the first steel city in 1869, formed the Soviet Union's industrial heart, powered by Europe's largest and most powerful hydroelectric dam, which later made possible the region's strategic aluminum industry. Other major industrialization projects took place at Magnitogorsk, the Kuznets Basin of Siberia, and Stalingrad.[74]

Similarly, Japan embraced state-sponsored industrialization. Because Japan lacked coking coal or iron, Japanese industrialization's environmental impact was spread across East Asia. Aware what the Opium War had done to China, the Japanese took alarm in 1853 when U.S. Commodore Matthew Perry arrived with warships and demanded that Japan open itself up to international trade. Emperor Meiji took power in 1868, dismantled the feudal system, and centralized power. When private capital failed to invest in industrialization, the government took over. Unable to control its own tariffs until 1911 due to unequal treaties, Japan exported silk and cotton textiles to pay for coal and iron ore to build up its military power. By 1914, over ten thousand miles of railroad lines knitted the mountainous islands together. A unique system arose in which family controlled holding companies ruled integrated systems of financial and monopolistic industrial firms. Industrial capacity and technical knowledge advanced quickly. Japanese victories over China in 1894–1895 and Russia in 1904–1905 set the stage for more aggressive Japanese foreign, imperial, and economic policies in the twentieth century.

Steel and environmental change

Carnegie's steel business illustrates the environmental consequences of every nation's iron and steel industry. Pittsburgh's environmental impact extended far beyond its smoke-filled valleys. For good quality iron, Carnegie reached over a thousand miles to the Mesabi Range of Minnesota. John D. Rockefeller had bought up distressed mines in the

Figure 8. John Vachon, World's largest open pit mine, Hull-Rust-Mahoning, near Hibbing, Minnesota, 1941. A hole in the ground two and a half miles long, three quarters of a mile wide, four hundred feet deep represents the great environmental price that the age of steam and steel demanded. (Library of Congress 2017813485. Farm Security Administration—Office of War Information Photograph Collection)

recently developed iron deposits of the Mesabi Range and got control of the railroad to Lake Superior and ore ships on the Great Lakes. In response, Carnegie bought mines and threatened to build his own railroad and shipping fleet. He and Rockefeller came to terms very favorable to Carnegie, to whom Mesabi iron ore now exclusively flowed.[75] The Mesabi Range paid a steep price. Its shallow ores favored open-pit mining. Open-pit mining reduced waste, since companies found it easy to map the various qualities of ore and combine low- and high-grades together for a salable product and less ore in tailings. Open-pit methods also allowed mining of ore that tunnel mines would have had to leave in the ground. Finally, rails could lead right into the pits and allow direct loading from steam shovel to rail car. However, the huge barren holes left by the method did huge violence to land, watersheds, and ecosystems.[76]

There also can be no steel without coal. The coal mines of western Pennsylvania that supplied power and coke to the mills produced air pollution, and acid drainage from mines and tailings severely damaged the drainage of the upper Ohio. Occasionally coal mines catch fire and burn hot but slow for decades, or perhaps centuries. Mines sometimes collapse, sometimes tragically with miners in them, and create sinkholes above them. In 1959, miners in the Knox Mine tunneled illegally under the Susquehanna River and the mine roof collapsed. Billions of gallons of water rushed in, drowning twelve miners, before the hole could be plugged.

Things go better with coke

Cooking coal for coke creates usable by-products like coal gas and coal tar. In England and by the end of the century in the U.S. and on the Continent, every major city built pipe networks to pipe gas to cleanly light streets, homes, galleries, and industry, and fire industrial processes like smelting iron and steel. Gas was convenient, clean, and safer than open fires.

Coal tar led to the modern chemical industry. Its chemical complexity fascinated early nineteenth-century chemists. The British textile industry was the catalyst. Factories produced far more textiles than such workers using natural products could process. They washed cloth in alkali detergents (processed from ashes) and soft water, "soured" it with lactic acid from sour milk to counteract the alkali, laid it in the sun to bleach, and starched and blued it.[77] Chemicals could be produced on demand and act far more rapidly than natural processes. James Watt's investor John Roebuck had made his fortune from developing a more efficient method to make sulfuric acid for bleaching. Chlorine extracted from salt in 1785 made an efficient, if noxious, bleaching agent. Scot Charles Tennant's chlorine bleaching powder of 1799 allowed bleaching without unpleasant fumes.[78]

Chemists investigating coal tar liberated textile processors from another natural product, organic dyes. Organic dyes, derived from various plants, woods, and one insect (cochineal), mostly came from distant tropical lands. Supply depended on successful harvests. Difficult, laborious processes produced dyes of unreliable quality and color. In 1856, while attempting to synthesize quinine from coal tar, London university student William Henry Perkin created the first aniline dye. Aniline dyes were cheaper, could be made in a greater range of colors, were easier to apply in industrial settings, and were far more colorfast than natural dyes. After Perkin successfully commercialized his discovery, chemical firms making synthetic dyes proliferated in England, France, the Rhineland, and Basel, among them such corporate giants as Ciba-Geigy, Sandoz, BASF, and Bayer. Because coal tar also had medicinal uses, many of these companies made pharmaceuticals and some later became pharmaceutical companies exclusively.[79]

The chemicals used to make dyes from coal tar produced noxious wastes in the air and water. In England, venting of fumes from the process created hydrochloric acid, which killed trees, hedges, and crops and injured animals and people downwind. Protest led to the Alkali Act of 1863. In Basel and Germany, dumping of effluents, particularly arsenic, into waterways raised opposition after a major poisoning incident in Basel in 1864 damaged fisheries. In their defense, factory owners asserted with truth that no direct evidence linked air pollution with human health and that humans had dumped waste into rivers since the dawn of civilization because rivers cleaned themselves as they flowed.[80]

Recognizing industrial capitalism's social and environmental price

Around 1870, as industrial capitalism entered the Second Industrial Revolution, the century of British leadership in innovation and mechanization which started with James Watt was ending. A century of dominance by Germany and then America commenced. Carnegie rode the crest of the wave into the twentieth century.

Carnegie's conscience did not let him rest to enjoy his fortune. Reformed Protestant culture imbued him with a moralistic view of the responsibilities of the wealthy to a righteous society. His own rise from poor weaver's son to powerful capitalist convinced him that self-discipline, self-education, and self-motivation led to success. Yet he knew full well the role that luck played in his life. As early as 1868, Carnegie jotted down private notes about wealth, calling its pursuit "the worst species of idolatry" and thinking about how to "spend the surplus each year for benevolent purposes."[81] In 1889, far wealthier and more prominent, Carnegie published the essay "Wealth," or "The Gospel of Wealth." He argued that the community more than the individual produced great private wealth. Hence the rich had a moral duty to use their fortunes for the betterment of society.[82] When Carnegie retired as the world's richest man, after selling his steel business to J. P. Morgan in 1901, he strove to give away his money. Carnegie built libraries, donated to universities, and created a relief fund for injured workers and survivors of victims of fatal accidents.[83] He threw himself into activism against imperialism and for peace. The World War, where advances in steel, technology, and chemistry caused horrific slaughter, was a brutal shock to him. He died in 1919. The world he had thrived in and helped to make was also passing away. Society and culture were changing. The steel business was changing. Capitalism itself was changing.[84]

FIVE

Conserving Resources

Industrial capitalism and limits to growth

In the mid-1860s, while Andrew Carnegie was making his first investments in the iron business, polymaths on opposite sides of the Atlantic published books that shook the confidence of a complacent public that the progress of industrial-capitalist civilization could be sustained. In 1864 in New York, Charles Scribner published *Man and Nature; or, Physical Geography as Modified by Human Action* by George P. Marsh. One year later in London, Macmillan and Co. brought out W. Stanley Jevons's *The Coal Question: An Enquiry Concerning the Progress of the Nation, and the Probable Exhaustion of Our Coal-Mines.* Humans of course had always extracted resources from soil and stone, albeit with increasing efficiency over the millennia, but Marsh and Jevons argued that industrial capitalism's unprecedented ability to extract natural resources threatened to deplete them so as to deprive future generations of their benefit. Worried conversations buzzed in the hallways of power, and anxious reports filled periodicals about what in the late twentieth century would be called limits to growth. For centuries, even millennia, thinkers had sounded alarms about society's destruction of one or another crucial resource, most often woodlands, but Marsh and Jevons for the first time made the compelling case that humans threatened to destroy the material bases for civilization, "like killing the goose to get the golden egg," said Jevons.[1] Marsh suggested solutions. Jevons suggested moral use of profits while they lasted.

Marsh's book had huge political consequences in the United States. *Man and Nature* rang alarm bells that called forth the American movement for conservation and forestry. Jevons's work prompted government commissions and further research and now is a classic of energy economics. The works of Marsh and Jevons heralded the birth of a new international movement, led with surprising regularity by men and women who, like them, were of Reformed Protestant or Protestant background. The movement for conservation of resources merged with other causes that arose in the 1850s and 1860s, including urban sanitation, pollution abatement, and urban and national parks, to produce twentieth-century environmentalism. Few conservationists, however, seriously questioned industrial capitalism itself. Rather, they sought to fix or ameliorate its problems and keep the goose healthy, happy, and laying golden eggs forever.

Man and Nature: *nature's ecological limits*

In *Man and Nature*, Marsh maintained that Americans and European colonists were repeating mistakes of antiquity. In travels around the Mediterranean, he had sought out places that ancient manuscripts described as lush, well-watered, and wooded, only to find bleak, barren rocks baking under the hot sun. A historical sequence laid itself out to him: inhabitants had cut the forests, set goats and sheep to eat grass off the hillsides, and watched soil erode and springs and streams dry up. "The operation of causes set in action by man," he observed, "has brought the face of the earth to a desolation almost as complete as that of the moon."[2]

Since the 1840s, Marsh had spoken and written on the rapid retreat of America's forests. If people neglected forests and soils, the United States and European colonies would look like those bleak Mediterranean landscapes. In the relatively brief period of European colonization, he warned, "great, and . . . sometimes irreparable, injury has been already done . . ."[3] People assumed that nature could recover from any hurt. Marsh shocked his readers with the idea that humans caused permanent destruction of nature's fertility and resources and warned that the decline signaled future poverty.[4]

Marsh understood how environmental change in America was tied to industrial capitalism. To feed the people and mills of Europe, American forests retreated to make pastures and fields for food, cotton, and tobacco.[5] Industrial demand decimated fish, mammals, and birds. Marsh noted that "all the processes of agriculture, and of mechanical and chemical industry" killed fish.[6] Globally, commerce and industry wantonly took what it needed and wasted the rest:

> The terrible destructiveness of man is remarkably exemplified in the chase of large mammalia and birds for single products, attended with the entire waste of enormous quantities of flesh, and of other parts of the animal, which are capable of valuable uses. The wild cattle of South America are slaughtered by millions for their hides and horns; the buffalo of North America for his skin or his tongue; the elephant, the walrus, and the narwhal for their tusks; the cetacea, and some other marine animals, for their oil and whalebone; the ostrich and other large birds, for their plumage . . . The great demand for oil and whalebone for mechanical and manufacturing purposes, in the present century, has stimulated the pursuit of the "hugest of living creatures" to such activity, that he has now almost wholly disappeared from many favorite fishing grounds, and in others is greatly diminished in numbers.[7]

Industrial capitalism was ravaging almost everything in the natural world it touched.

The Coal Question: *nature's limits to growth*

In contrast with Marsh's expansive global analysis, Jevons's investigation extended only to British coal reserves. Yet he, too, concluded that humans threatened to exhaust natural resources. He noted "that the Coal we happily possess in excellent quality and abundance is the mainspring of modem material civilization." Its heat drove engines and industrial processes, warmed hearths, and cooked food. Its chemistry yielded a variety of colorful dyes and useful substances.[8] All that made Britain great and powerful—its iron and steel industry, manufacturing, industrial technology, transportation, chemical industry, and imperial might—derived from practically inexhaustible coal seams beneath the green and pleasant land. Jevons argued, however, that when easily mined coal was gone, the expense of mining would increase. Other nations with large, accessible coal supplies—Jevons repeatedly mentioned Pennsylvania—would have an economic advantage. Britain would regress to a pre-industrial economy.

Could reduced waste and improved efficiency extend the age of national greatness? No, argued Jevons. Greater efficiency reduces the cost of burning coal and leads to greater consumption of coal, not less. James Watt's steam engine was far more efficient than a Savery or Newcomen engine.

Lower fuel costs freed capital for business expansion and additional steam engines burning additional coal. Greater efficiency also encouraged steam power for new purposes, such as railroads and steamships. These factors, multiplied as well by population growth and rising national income, caused demand for coal to escalate dramatically.[9] Jevons added that, since any nation could also use efficient engines, improved efficiency gave Britain no competitive advantage. The principle that greater energy efficiency leads to greater energy consumption is called the Jevons paradox.

Great Britain, Jevons predicted, was approaching what today might be called "peak coal." Within perhaps a century, the era of British power and prosperity would end. "We are now in the full morning of our national prosperity," he warned, "and are approaching noon."[10] Britain had no great supplies of any potential substitute. Considering the alternatives of burning quickly through the nation's coal supplies or moderating energy use to make coal last, he concluded, "*We have to make the momentous choice between brief greatness and longer continued mediocrity*"[11]—burn out or fade away.

The Coal Question, by a young, unknown author, provoked a national scare. John Stuart Mill quoted it on the floor of Parliament. Chancellor of the Exchequer and future prime minister William Gladstone read it and corresponded with Jevons. Parliament set up a national commission to investigate Jevons's claims. Its conclusion that Britain would remain the world leader in coal production for the foreseeable future calmed fears. *The Coal Question* fell from favor, until the day came a few decades later when Jevons's prediction came true.[12]

Two paths to the question of resource limits

Although these environmental Jeremiahs both warned of resource dearth and decline of civilizations, Marsh and Jevons came from very different worlds that shaped their critiques in very different ways. Born in 1801 in Woodstock, Vermont, Marsh grew up in the strict but cultured home of a Dartmouth-educated lawyer and Federalist politician. His grandfather had served in the French and Indian War in Connecticut and moved to Vermont in 1772. A militia colonel during the Revolution, he held offices of lieutenant governor and judge after the war. Marsh also graduated from Dartmouth College and practiced law. In 1843, he was elected Whig representative to Congress, where he helped create the Smithsonian Institution.

In Vermont, Marsh saw how the Industrial Revolution wrought provincial environmental devastation akin to that in the Mediterranean. Farmers cut down and burned trees to sell the ashes to local potash factories. English manufacturers imported the potash to render into fuller soap to wash wool for manufacturing yarn. For steadier income, farmers introduced Merino sheep in 1810 to graze the green but now treeless slopes and sold their wool to New England textile mills. By 1825, half a million sheep were cropping Vermont's hillsides bare. Rains ran unimpeded off naked slopes, flooding streams and rivers and washing out bridges and mills. Erosion silted streams and spoiled fish and bird habitat. On dry summer days, treeless hillsides baked in the sun. Springs and streams dried up.[13] Too reliant on extractive industries, Vermont's future looked bleak.

Marsh worried that settlers in the new western states were following the same path to environmental degradation and poverty. Since 1847, he had advocated better husbandry of the soil and better care for woodlands. While minister to the Ottoman Empire, he traveled to inspect Mediterranean soils and agriculture. Returning to Vermont in 1854, he continued public service even as failed investments in a corrupt Vermont railroad bankrupted him. He authored a state report on fish hatcheries and served as state railroad commissioner. Marsh was appointed ambassador to Italy in 1861, a post he held until his death in 1882. Light duties gave him time to distill his observations from Vermont and the Mediterranean into *Man and Nature.*[14]

By contrast, William Stanley Jevons was born and raised in the grimy urban heart of the world of coal and iron. His grandfather, a nailmaker in the Black Country near Birmingham, came to Liverpool in 1798 as agent for a nail firm. He became a prosperous independent iron merchant. Jevons was born in Liverpool in 1835. Like Marsh, he grew up in a cultured and well-to-do family. Financial tragedy followed the bursting of a railroad bubble. Railroad construction halted and demand for iron collapsed. Jevons's family business went bankrupt in 1848. Declining family finances interrupted Jevons's studies at University College in London. In 1853, he took a job in Australia as assayer in the Royal Mint, established after the 1851 gold rush. Abundant free time allowed Jevons to investigate and publish on a variety of subjects, from railway policy to meteorology to urban sanitation in Sydney. In 1859, he returned to complete his B.A. and M.A. at University College in London and was hired at Owens College in Manchester.

In a quest to put economics on a firm mathematical basis and thereby explain business cycles such as the one that destroyed the family business, Jevons published important works in the fields of logic, economics, and scientific theory. In *The Coal Question*, Jevons brought statistics and mathematics to bear on the hotly debated topic of whether Britain's coal supply set limits to economic growth. In 1876, he was appointed Professor of Political Economy at University College in London. Five years later, he resigned to devote himself to a great work on economics. Jevons desired to create a rational economics, a kind of liberal or libertarian economic machine that would run without outside intervention. An accidental drowning while swimming cut his life short in 1882 at age forty-six, three weeks after Marsh died.[15]

Puritan roots of limits of natural resources as a social and moral question

Yet, despite these differences, Marsh and Jevons shared a Puritan heritage that shaped their analyses and solutions. Marsh's family was staunch Congregational Calvinist, although he himself drifted from orthodox Calvinist belief. "The last of the Puritans," as his wife called him, he neither embraced nor disavowed the religion of his forefathers.[16] Jevons's Puritan Presbyterian ancestors had exchanged Calvinism for Unitarianism. Jevons had Unitarian ministers on both sides of the family. As a non-Anglican shut out of Cambridge and Oxford, he earned his degrees at secular University College and taught at secular Owens and then University College. Although quite religious, he rarely attended church in later life.[17]

Marsh's ideals and values bore the stamp of the ambition of Puritan founders of New England to build a godly, equitable, just, and sustainable society. To that end, colonial governments granted land to towns, not individuals. Towns divided up land equitably (although not equally) to banish poverty. Ministers preached the importance of suppression of the self for the common good and the moral duty to be sober, industrious, and useful to others. Puritans shared with

Presbyterians a strong ethic of stewardship, the Calvinist duty to God to improve one's possessions. Farmers practiced a type of sustainable agriculture. Because trees were essential for fuel and timber, New England towns and colonies regulated farmers' wood lots to ensure resources for future generations.[18]

Moreover, Reformed theology promoted a spiritual view of nature, the religious value of the natural sciences, and proto-ecological thinking.[19] The natural world was a theater for God's glory. Calvinism taught a profound version of providence in which God directs the smallest molecule in an intricately interconnected universe. All nature shows the wisdom, goodness, and existence of God, and is where God draws closest and communicates himself.[20]

Puritan moralism, ecological thinking, and practicality pervaded *Man and Nature* and lent passages a sermonic character. To Marsh, humans had broken the web of creation that connected the elements of nature or, in modern terms, disrupted ecosystems. "Man is everywhere a disturbing agent. Wherever he plants his foot, the harmonies of nature are turned to discords. The proportions and accommodations which insured the stability of existing arrangements are overthrown. Indigenous vegetable and animal species are extirpated, and supplanted by others of foreign origin, spontaneous production is forbidden or restricted, and the face of the earth is either laid bare or covered with a new and reluctant growth of vegetable forms, and with alien tribes of animal life."[21]

Marsh did not lay responsibility at the feet of industrial capitalism per se (the pre-industrial Mediterranean had after all suffered similar problems), although he excoriated unscrupulous corporations for causing "the decay of commercial morality" and "corruption" of government and justice.[22] Instead, Marsh blamed an individual moral lack of responsibility and decried "the debt which the prodigality and the thriftlessness of former generations have imposed upon their successors," who now face "the necessity of restoring the disturbed harmonies of nature, . . . thus fulfilling the command of religion and of practical wisdom, to use this world as not abusing it."[23] "Man has too long forgotten that the earth was given to him for usufruct alone, not for consumption, still less for profligate waste," he preached.[24] "Man" had been a bad steward.

Marsh argued that we must reclaim "lands laid waste by human improvidence or malice" and become "a co-worker with nature in the reconstruction of the damaged fabric which the negligence or the wantonness of former lodgers has rendered untenantable."[25] How was this to be done? Marsh idealized the New England town as the model of a democratic, moral community of self-restrained stewards of the land. Only one paragraph in his long, compendious work discussed the potential good of government works, albeit scattered passages recognized the work of restoration or forest protection that municipal, local, and national governments had done. As for the United States, Marsh commented repeatedly on the weakness of state and national governments to protect and preserve public lands, making it individuals' responsibility to practice conservation.[26]

The Coal Question was more of a Unitarian appeal to reason than Congregational evangelical sermon. Whereas the epigraph of *Man and Nature* quoted Connecticut Congregational minister Horace Bushnell, that of *The Coal Question* cited Adam Smith. Under Enlightenment influence, English Presbyterianism had evolved into a Unitarianism that blended rationalism and Puritan moralism. As a Unitarian, Jevons valued the power of reason, believed in one's duty to morally improve the world, sympathized with outsiders, and respected individual conscience. His work in logic and economics reflected his interest in social questions. A student in London between 1851 and 1853, Jevons took long walks through London's poor neighborhoods and manufacturing districts. Outbreaks of cholera led him to read Charles Dickens's calls for sanitary and social reform, commonly regarded as related issues. He turned to social questions late in life. His publications included *The State in Relation to Labor* of 1882 and the posthumous *Methods of Social Reform* of 1883. As an urban Unitarian, Jevons gave much greater attention to social problems than Marsh.[27]

Major contemporary liberal and utilitarian economists shared Jevons's Dissenter roots or Unitarian beliefs. British free-market economics grew rather comfortably out of Unitarian ideals and cultures of religious outsiderness. Most Unitarians came from Dissenter families, but many Anglicans and Jews embraced Unitarians' regard for individual conscience, cool rationality, and universality. Among Jevons's major economic predecessors and peers numbered Adam Smith, professor of moral philosophy at the University of Edinburgh and Presbyterian attracted to rational religion; Huguenot economist and businessman

Jean-Baptiste Say; Utilitarian philosopher Jeremy Bentham, who derived his philosophical standard of "the greatest good for the greatest number" from the writings of Unitarian minister Joseph Priestly; David Ricardo, child of Sephardic Jewish immigrants from the Netherlands and convert to Unitarianism; and John Stuart Mill, child of a Scottish Presbyterian-minister-turned-rationalist.

Moral concern shone clearly in passages of *The Coal Question*. In the introduction to the second edition, Jevons discussed Britons' moral duty to themselves and posterity while their coal advantage was at its zenith. He commented "that the whole structure of our wealth and refined civilization is built upon a basis of ignorance and pauperism and vice. But we are now under a fearful responsibility that, in the full fruition of the wealth and power which free trade and the lavish use of our resources are conferring upon us, we should not omit any practicable remedy. If we allow this period to pass without far more extensive and systematic exertions than we are now making, we shall suffer just retribution." He added, "We ought not to think of interfering with the free use of the material wealth which Providence has placed at our disposal, but that our duties wholly consist in the earnest and wise application of it. We may spend it on the one hand in increased luxury and ostentation and corruption, and we shall be blamed. We may spend it on the other hand in raising the social and moral condition of the people, and in reducing the burdens of future generations. Even if our successors be less happily placed than ourselves they will not then blame us." Jevons concluded with the reflection that, as national prosperity was approaching its noontime, "we have hardly begun to pay the moral and the social debts to millions of our countrymen which we must pay before the evening."[28] With great coal power comes great social responsibility.

Conservation before Man and Nature *and* The Coal Question: *the soil*

Although Marsh and Jevons sparked a major movement for conservation of natural resources, concern for declining resources already had a long history. Agriculturalists in particularly had long contended with declining soil fertility. Everywhere around the world, local farmers developed methods to maintain or restore soils—if, that is, they had ownership or control of their land. Owners of large estates or absentee landowners often required production of certain foodstuffs or certain quantities, disincentives to soil conservation.

A true science of agriculture arose in Europe in the sixteenth century. Renaissance interest in agricultural works of antiquity set off a vogue in agricultural manuals for the gentleman or noble farmer, mostly rehashes of the classics. In the sixteenth century, Huguenot authors began writing agricultural handbooks based increasingly on science and experience, such as books by Bernard Palissy in 1563 and 1580, Charles Estienne and Jean Liébault in 1572, and Olivier de Serres in 1600.[29] Huguenot agricultural science crossed the channel when Sir Hugh Plat in 1594 published the first work of English agricultural chemistry, which discussed Palissy. At the same time, "improving" the soil became a moral act in English Calvinism. "Improvement" had connoted management of something for profit until Calvinist cleric Samuel Gardiner gave the term a religious spin in 1606.[30] Improvement became a catchword for one's moral duty to make oneself and one's possessions better. The term came to mean specifically greater agricultural productivity through experiment and science. It was an era of great optimism for ending dearth and making waste places produce, although improvers disregarded traditional practices and knowledge and justified enclosure of commons.[31] Then, in 1798, Thomas Robert Malthus, an Anglican cleric educated at a Lancashire Dissenting academy and tutored by a Unitarian, published *An Essay on the Principle of Population.* It predicted that population would tend to outrun food supplies and asserted that nature would limit the growth of population.[32] Could food supplies keep up? The improvers had not yet shown that they could.

In answer to Malthus, the sciences of chemistry and botany bounded forward. After Alexander von Humboldt reported how Peruvian farmers used guano from seabirds to make barren fields yield, Humphrey Davy investigated guano's properties and published an 1813 book that described how plants needed certain chemicals to thrive, particularly the nitrogen and phosphorus that guano possessed in abundance. In Germany, agricultural chemist Justus Liebig in 1840 published a widely translated and influential book that singled out nitrogen, potassium, and phosphorus as essential elements for plant growth. Germans now led the field of scientific agriculture. They established the first publicly funded agricultural research station, in Saxony, in 1852, which inspired many more in Germany, Austria, Italy, Japan, and the United States.[33]

Davy and Liebig incited a guano rush in the Pacific, where enormous amounts covered isolated islands, home only to birds for millennia. Workers labored in slavery or near slavery in clouds of ammonia fumes to mine the fossil excrement. Britain, the Netherlands, Germany, and the United States imported it. For American plantation owners, guano seemed to solve their unsustainable agricultural practices. Farmers and planters came to rely on an additive to soil that must be purchased and with which they could not supply themselves. Such external inputs, combined with mechanization, foreshadowed the rise of factory farming.[34]

Science, conservation, and the agricultural crisis of industrial capitalism

Marsh's desire to save the farmer was widely shared in both Europe and America. Agricultural improvement had begun to chase its tail. The ancient drive to increase the product of the earth to feed more people led farmers deep into the market exchange system of nineteenth-century capitalism. Ever greater production required ever greater capital investment. Greater production pushed prices down, which encouraged ever greater production to pay for investments and make a profit, which put unsustainable demands on soil.

Britain led the way. The envy of and model for the United States and Continental countries, British farmers had attained quite high production by 1850. To achieve this landmark, they enclosed commons into more efficient private plots, improved stock, developed new techniques to drain land, and added large amounts of imported guano, nitrate (for nitrogen), gypsum, lime, phosphates, and potash (for potassium). It was highly productive, yes, but so capital intensive that expensive British food required tariff protections. Repeal of the Corn Laws in 1848 exposed British agriculture to international competition. Ever-faster ships brought a glut of American and Canadian grain to join Russian wheat and drive down prices, leading to a collapse of wheat prices in the 1870s. Chicago's mass-produced flour and meat products also affected global markets. Development of refrigerated transport allowed meat and dairy products from the Americas and Australasia to pour out across the world. The

British rural landscape changed. As farming grew more mechanized and farms larger, labor demand declined. Villages emptied out. With beef more profitable than grain, pasture expanded as arable shrank.[35]

American farmers on older lands also struggled. When Indian removal opened up western lands in New York and the Great Lakes region, New England farmers moved west and shipped abundant streams of grain back east. Farmers on old, worn eastern soils specialized in truck gardening or opened the woods and fields to dairy farms for growing industrial cities. After the Civil War, transcontinental railroads brought thousands of farmers to the Great Plains. Railroads conveyed their produce to Chicago millers. The technology of farm machinery advanced year by year. Farming large, flat fields where labor was scarce grew easier but enticed farmers into debt. Mechanization improved thin profit margins but higher production inevitably pushed commodity prices down. Cheaper food meant better-fed people but it also indebted farmers and drove people off the land in eastern states as in Britain.

Southern farmers and planters also were in trouble. After the Civil War, sharecropping and tenantry replaced the plantation system and tied black farmers to specific plots of land. No longer could farmers abandon exhausted fields and move to freshly cleared ones. Landlords required sharecroppers and tenants to raise cotton, rather than rotating with less-remunerative crops or fallow. Yields declined, even while prices were dropping. Farmers responded by planting more cotton. White farmers fell into debt and joined the ranks of tenants and sharecroppers. Southern poverty deepened and exhausted Southern soil eroded away.[36]

Farmers found various ways to respond. On the European Continent, where farmers were numerous and the agricultural sector more important to economies, most nations responded to farmers' plight with protective tariffs. English and Americans founded agricultural journals and formed local agricultural societies, which gave farmers the latest advice and agricultural science. Britain and the U.S. invested in research and education. Britain established its first agricultural chemical laboratory in 1842 in Edinburgh and organized educational institutions to inform farmers or their sons of the latest science.[37] In America, New Englanders and their western kinsmen sought to support agriculture through scientific and professional institutions. They established professorships of agricultural science. They lobbied for creation of the Department of Agriculture in

1861 and the Division of Forestry. They campaigned for state land-grant colleges geared to mechanical and agricultural education. They set up the first American agricultural experiment station in Connecticut, a model for others established under state and national auspices.[38]

Conservation before Man and Nature *and* The Coal Question: *timber and fuel*

Marsh's concern for forest conservation expressed another ancient worry. Deforestation had alarmed authors and authorities at many times and places. Pre-industrial Japanese, Chinese, and Indian rulers dealt with the issue with widely varying methods and success. Local control of forests preserved wood for construction and fuel. Rulers preserved forests to prevent erosion and maintain sources of timber.[39] European aristocrats preserved woods as habitat for game.[40]

Military concern was a strong motivation for state forestry. Sixteenth-century Venice set up government forests, since its power and wealth depended on a strong navy and merchant marine.[41] To supply their navies and merchant ships in the Mediterranean and Red Seas and in the Indian Ocean, the Ottoman Turks tightly supervised timber use, cutting, and shipments in forests on the coasts of Anatolia and the Black Sea.[42] Spain's powerful navy relied on forests in northern Spain. Charles V issued the first edict in 1547 to conserve and expand forests for the navy. Philip II bureaucratized Spanish forestry after heavy loss of ships in the defeat of the Armada in 1588.[43] Eventually, however, Spain and the other European powers outsourced timber needs. In the eighteenth century, Spain built ships in forest-rich Cuba, while Portugal turned to Brazil. Much of England's naval timber came from the Baltic, North America, and India. The Dutch, with almost no forests of their own, imported timber from the Baltic and Germany and teak from the Dutch East Indies.[44]

By the late seventeenth century, concerns about securing adequate supplies of timber for navies inspired two landmarks of forestry: John Evelyn's *Sylva: or, A Discourse on Forest Trees* of 1664, and the *Ordonnance* of 1669 of Jean-Baptiste Colbert, minister to Louis XIV of France. England's lack of extensive royal forest land and the rights of its landowners prevented Evelyn's proposals from taking effect.[45] Initially ineffective, Colbert's Ordinance set France on a path that would win French forestry international respect by laying the basis for later forestry law. During the French Revolution, the state took possession of the forests of the Church and nobles, to which French governments applied forestry methods and rules. In 1824, France established an influential forestry school at Nancy.[46]

In the nineteenth century, Germany became the other leader in scientific forestry. In the aftermath of destructive wars, German states sought to inventory and rationally manage their resources, including forests. Local control seemed too unsystematic and inefficient to maximize economic returns. In rationalized German forests, sylvan monocultures stood in rows like Prussian soldiers on parade, to be harvested and replanted with a predictable sustained yield. Other nations hired German foresters for their own forests and forestry schools.[47]

Forestry in British India and the United States

The way was prepared for the enthusiastic reception of *Man and Nature.* The British Empire and the U.S. both faced ecological problems that settlers and private corporations, ignorant of local land management practices, caused in new and unfamiliar landscapes. The issue showed up first on islands, where colonists rapidly ran through their rich but limited forests, but it was British officials in India, not on islands, who established a model of scientific forestry. A network of botanists and surgeons in the colonies raised the alarm over deforestation and its consequences of desiccation, floods, and erosion. Scots would play a critical role, thanks to the medical school of the University of Edinburgh. Edinburgh had modeled its medical curriculum on the University of Leiden's program, in which natural history formed an important part of the course of study. Edinburgh-trained surgeons flooded into the army, navy, and colonial service and formed a network of natural scientists. Scots and Edinburgh graduates founded many of the empire's early botanical gardens.[48]

Scots and Edinburgh graduates also led the campaign that produced the Indian Forestry Service. Scottish doctor and influential botanist Hugh Cleghorn grew concerned about Indian deforestation's effect on river flow and desiccation. Around 1850, his report on destruction of Indian forests caught the eye of the Earl of Dalhousie, Scottish governor-general of India from 1848 to 1856, whose actions facilitated creation of the Indian Forest Department. Out of Cleghorn's activism and advocacy, the British government, which had taken over the administration of India from the East India Company after 1857, established an all-India Forest Department in 1864, the year Marsh published *Man and Nature.*

Cleghorn recruited German Dietrich Brandis to join him on the Commission of the Conservancy of Forests. Brandis staffed leadership positions with graduates of French and German programs, who brought their professional distrust of traditional peasant control of forests. State-sponsored scientific forestry for sustained yield had come to the Empire and steamrollered its way over local tradition and practices, just as it had in France and Germany.

And yet, not quite as it had in Europe. Brandis, Cleghorn, and many other Indian foresters read *Man and Nature* and began to understand the ecological importance of Indian forests alongside their commercial and other uses. Cleghorn and Brandis both corresponded with Marsh. "I have carried your book with me along the slope of the Northern Himalaya, and into Kashmir and Tibet," Cleghorn wrote him.[49]

Along with Marsh, another factor attenuated German–French influence on Indian forestry. The leadership of the India Forest Service was Continental, but the rank and file was mostly Scottish. Scots brought a multiple-use version of forestry, along with a distrust of the greed and self-interest of commercial interests, which they sought to exclude from forests. Scottish forestry began in the seventeenth century when large Scottish landowners planted forests to improve their lands. Forests served many purposes beyond timber. Trees beautified estates. They created habitat for game. Owners preserved coppiced woods and experimented with non-native trees, planting millions over the bare hills of the Highlands. Smaller landowners aped the larger. A cadre of experienced foresters grew. After a civil service exam was instituted for the Indian Forestry Service, the proportion of Scots dropped. Nevertheless, aspiring British foresters took the forestry course at Nancy and then worked under an approved forester on a Scottish estate before heading to India. With the need to adapt to very diverse forest conditions, India's foresters developed a hybrid forestry. Brandis, too, knew well that forestry could not succeed with the opposition of local people. Regimented German- or French-style forest plantations rarely appeared in India.[50]

Marsh's *Man and Nature* also inspired those who dominated the development of professional forestry in the United States, almost all of whom were like him no more than a generation away from a New England Congregational Church. Franklin B. Hough noted the alarming drop in acreage of forest land in New York, while Marsh's book confirmed and reinforced his sense of crisis.[51] In 1880 he was appointed the nation's first chief of the new Division of Forestry in the Department of Agriculture.[52] Nathaniel Egleston, an ordained Congregational minister, succeeded him in 1883.[53] Prussian forester Bernhard Fernow succeeded Egleston in 1886, followed a dozen years later by Gifford Pinchot. Pinchot oversaw the agency's transformation to the Bureau of Forestry in 1901 and, with transfer of forest reserves to his control, to the Forest Service in 1905.[54]

American foresters followed the lead of British Imperial Forestry and the French and Germans, adapted to American conditions. Pinchot staffed his new Forest Service with foresters trained in new American forestry schools. Professional forestry now had German, French, Scottish, Indian, and American variants, distinctive in form and detail. They shared the influence of the Prussian ideal of efficient, scientific forestry under central government management for profit and the common good, but evolved into looser, more ecologically informed versions serving wider purposes.[55]

Rise of wildlife conservation

Marsh's concern about capitalist societies' wanton destruction of wildlife was widely shared. The rapid disappearance of buffalo and passenger pigeons shocked Americans, already aware of declining abundance of wildlife. With great efficiency, market hunters delivered huge amounts of game to local and urban grocers. Other hunters sought birds with valuable plumes to adorn the hats of fashionable ladies, nearly destroying many species of large birds to do so. Sport hunters worried that they would have little left to shoot and campaigned to suppress market hunting and regulate sport hunting. After the turn of the century, states began to pass hunting regulations. British sport hunting and game collecting in South Africa also sparked alarm. In the twentieth century, the British created game preserves throughout their east African colonies, most of which became national parks. Britain instituted game laws that tended to segregate whites, blacks, and game. In both the U.S. and the British Empire, enforcement was underfunded and often ineffective, but the laws set precedents. The legacy in Africa was complex, entangling wildlife conservation with glorification of the hunter and the hunt, suspicion of local peoples, and imperialism.[56]

Relieving the social and environmental crises of the industrial city

Marsh and Jevons worried about industrial capitalism's voracious appetite for natural resources, but neither remarked on how the tide of resources flowing in produced a stream of waste going out. The public, however, grew disturbed at poisoned air and water and unlivable cities. A coalition evolved that would join conservation and parks advocates in the environmental movement of the twentieth century.

Cities had always been dirty and prone to disease and epidemics, and Londoners had been breathing coal smoke for centuries,[57] but industrial cities marked a new, dark era. Manchester acquired a well-earned reputation as the dirtiest, unhealthiest city in Britain. It sat on a river at the head of an estuary at the center of a semicircular mountain range, the Pennines, one of the wettest regions of England. The descent of that moisture to the sea offered many excellent sites for waterpower and provided clean, soft water for processing and dying textiles. Nearby mines sent coal to Manchester by England's first modern canals. As coal-powered steam engines proliferated, heavy black smoke filled the natural amphitheater of the Pennines. Dense fog formed around coal particles. Soot and smoke blackened the rains. Gaslight relieved the gloom and slightly decreased the need for coal fires, but gas was a by-product of the smoky process of cooking coal for coke. Coke oven operators dumped the by-product coal tar (useless until mid-century) into rivers, where it mixed with ash cleaned out of steam engines to clog and befoul the water.[58]

Workers lived in the murkiest, grimiest districts within walking distance of factories, which crowded near waterways for power, transportation, and water for industrial processes. Private speculators put up shoddy, airless tenements. Black coal smoke and fumes from industrial processes deprived residents of clean air. So many people in close quarters presented a sanitary nightmare. City government struggled to provide sufficient (and sufficiently clean) water. Organic waste also presented challenges. Human waste accumulated in cesspits and privy vaults. A city-owned service collected it regularly if infrequently, which the city sold for a profit to farmers for fertilizer and avoided expensive sewage systems. Cities could not function without horses pulling wagons of freight, food, goods, and animal fodder. Horses dropped prodigious quantities of urine and manure on the streets, and in death remained on the street where they collapsed until city services removed them and dumped them into rivers. Disease was rampant. Manchester's death and infant mortality rates were several times the national average. Tuberculosis, a disease of crowded dwellings, raged. Epidemics swept through regularly. Cholera, a water-borne disease out of India, reached Britain in 1832 and killed thousands.

People who could afford to do so moved to the cleaner, healthier outer parts of the city and had indoor plumbing and water closets. Out of motives of money and moralism, the city designed it this way. Early in the century, Unitarian Liberals took over the city government and ruled until the arrival of universal manhood suffrage in 1881. To keep taxes low, they sold night soil (human excrement) to farmers to pay for waterworks but needed to keep indoor plumbing and water closets out of workers' districts so that their excrement could be collected. They ascribed poverty, disease, and filth to the poor habits and hygiene of Irish Catholic workers.

Appalled visitors to new industrial cities condemned black, smoke-filled air and foul, grimy, crowded workers' districts that contrasted so shockingly with the pleasant middle-class neighborhoods and beautiful mansions of factory owners at the edge of town. It took time, however, for local authorities to realize that these were problems that required action to correct. Even then, doubts and disagreement about what problems to address, how to identify the causes, and how to remedy them delayed action for decades. With the spread of industrial capitalism to other nations, their citizens, too, found themselves inundated with human, animal, and industrial waste, gasping for air, and gagging on bad water. Americans shared Britons' distress at urban poverty and disease and likewise blamed the moral failings of immigrants.

Blame for disease shifted slowly from the moral character of individuals to environmental causes, which doctors and engineers could more easily address. Theories that "miasma" or bad smells caused disease led cities to respond with measures to drain malarial marshes and clean away dead horses, rotting food, and human wastes. Realization dawned that clean water somehow was key. Free-market liberalism's ascendency prevented

English cities from raising funds for major infrastructure projects, so they relied on private water companies. Disease outbreaks and epidemics frightened cities into taking over the job of supplying good water. In the U.S., many cities accessed water upstream or built municipal reservoirs, aqueducts, and water works to supply clean, healthful water. Reservoirs needed to be remote enough to be clean but near enough to easily pipe to the city, which led cities to buy up and inundate villages and farmland. For cities whose local water was foul or turbid, the Scots invented the first municipal sand filtration system in Paisley in 1804, while three years later Glasgow installed the first water filtration system connected to city distribution pipes.

London took the lead in England in 1855, when Dr. John Snow meticulously demonstrated that cholera victims in a London neighborhood all drank from the same befouled well. Authorities shut down polluted wells and the city forced water companies that delivered water from the polluted tidal stretches of the Thames to draw from cleaner water upstream.[59] A turning point came by 1840 with Edwin Chadwick's studies in public health. His "Sanitary Idea" proposed piped delivery of clean, pure water to every house, complemented by water closets connected to sewers. The Sanitary Idea influenced public health authorities around the world. Within a hundred years, all major modern cities supplied water to homes and businesses and carried away human waste in sewers.[60]

What to do with the waste posed yet another problem. The work of Louis Pasteur and Robert Koch convinced medical authorities of the germ theory. By the end of the century, sanitary engineers also realized that the germ theory suggested ways to treat sewage before dumping it into waterways, but the expense and the awareness that benefits accrued only to non-ratepayers downstream slowed acceptance of the technology. In some cities, the early decision to save money by combining storm and sanitary sewers into one system would make it difficult to treat sewage because significant precipitation could overwhelm the system.[61]

The industrial city and birth of the parks movement

The grimness of industrial cities inspired a movement to provide workers with access to nature and clean air. Those with means traveled and vacationed in rural or natural areas outside cities. Technology brought nature closer. Steamboats and railroads reached into regions remote from crowded cities and in America promoted nature tourism's wild growth in popularity.[62] The early nineteenth century witnessed efforts to rethink and redesign the city to make urban life more pleasant and healthier. Parks offered a way to let in clean air, provide space for recreation, and provide uplifting moral influences of God's natural world. Britain's first urban parks began as royal parks that their owners gradually opened to upper and then middle classes. In the 1830s, nature became a selling point. A developer built Regent's Park to raise the value of surrounding plots, whose sale paid for the park. Other developers copied the strategy, as in Prince's Park in Liverpool. Initially restricted to those who purchased nearby houses, the parks opened to the public when public funds replaced private subscriptions for maintenance. In 1829, Scottish landscape gardener John Claudius Loudon began a campaign for publicly funded parks that culminated in creation in 1844 of Victoria Park in the crowded slums of East London. Concerned industrialists donated lands for the betterment of workers. Other Victoria Parks sprang up around England.[63]

The town of Birkenhead, a commuter suburb across the Mersey from Liverpool, built the first fully publicly funded park. The city hired Joseph Paxton, the architect who designed Prince's Park. Paxton turned 125 acres of unattractive, swampy, low-lying land at the foot of a sandy hill into a little Eden of woods, glades, winding paths, and scenic drives. Free to the public, the park opened in 1847. The idea of publicly funded parks spread across Europe. In 1850, a former Connecticut Congregationalist, Frederick Law Olmsted, visited Birkenhead Park. Returning to the U.S., in 1853 he entered a design inspired by Birkenhead in a design competition for a public park in New York City. It won and became the basis for Central Park, which inspired dozens of other American public parks in cities from Boston to San Francisco.

Parks were not the only way to bring nature to the people. Factory owners established ideal communities with plentiful greenspace. Owners of early British mills had to build not only the mills in generally remote locations near falling water, but also residences, schools, churches, stores, and workshops for millworkers and service workers. Towns had to be pleasing enough to attract workers. Many owners were Dissenters who sought to create happy, harmonious, and moral communities around their mills. Hence David Dale built a model milltown in New Lanark, Scotland, well known by the 1790s as Britain's largest spinning concern run with enlightened paternalism. Influenced by Manchester Unitarians, Robert Owen bought New Lanark in 1799, where his efforts to raise workers' condition through education and retraining made him influential in British socialism.[64] A different approach inspired the Protestant industrialists of Mulhouse in Alsace. In 1853 they formed the Société mulhousienne des cités ouvrières to build a model industrial city. To encourage thrift, workers did not rent but bought their housing units on a time plan at cost. Based on architect Henry Roberts's influential *Dwellings of the Labouring Classes* of 1850, each of four units in each quadruplex had its own garden on two sides, a bit of country in the city. The Mulhouse model inspired imitators throughout western Europe.[65] Many American industrialists also built model company towns.[66] For the middle and upper classes, improving urban transportation encouraged suburbs with lots of green. One of the earliest and most influential in America was Olmsted's design for Riverside, near Chicago.[67]

Around the turn of the twentieth century, city planners began to think more holistically. Sociology suggested that planned design of the city environment might solve social problems. The City Beautiful movement in the United States and the Garden City and town planning movements of Great Britain aimed to refine industrial cities into uplifting and improving environments and to restore the harmony of individuals, society, and nature. The notion that the planned city of the industrial age required new thinking grew extremely popular from America to Europe to Japan.[68]

Soon parks left their urban settings behind. Marsh had envisioned a sort of park whose purpose foreshadowed that of national parks: ". . . Some large and easily accessible region of American soil should remain, as far as possible, in its primitive condition, at once a museum for the instruction of the student, a garden for the recreation of the lover of nature, and an asylum where indigenous tree, and humble plant that loves the shade, and fish and fowl and four-footed beast, may dwell and perpetuate their kind, in the enjoyment of such imperfect protection as the laws of a people jealous of restraint can afford them."[69] By coincidence, in 1864, the year *Man and Nature* was published, the Federal government established Yosemite in California, the first park for the purpose of preserving natural beauty. *Man and Nature* more directly inspired the creation of huge Adirondack State Park between 1875 and 1895 to preserve both its beauty and its vital watershed.[70] In 1872, the U.S. created the world's first national park, Yellowstone. The national-park idea quickly caught on in Protestant nations, followed eventually by parks in most nations. National parks appeared around the British Empire. Switzerland created Europe's first, but the scarcity of sparsely inhabited landscapes forced European nations to create fewer parks or to create parks that included towns and cultural sites as well as natural beauty.

Conservation

During the tumultuous era between the 1880s and the mid-1910s, citizens, industrialists, and governments attempted to solve problems that attended industrial capitalism's global triumph. The offensive sight of industrialists enjoying great wealth while their workers lived amid filth, squalor, and terrible pollution provoked socialist, anarchist, and revolutionary solutions and political movements. Middle and upper classes hoped that bettering the condition of the poor would bring improvements and prevent radicalization. Germany and other countries passed social welfare laws to undercut the socialist movement. Pollution, however, was a more intractable challenge. Smoke Abatement Leagues appeared in mid-century in Britain and later the United States. In Germany, rising dissatisfaction with smoky cities inspired few citizens' groups and little political action, due to misplaced faith in strong and effective governmental response. Roadblocks that barred the way to cleaner skies included the technical difficulty and expense of reducing smoke; uncertain medical evidence of smoke's danger to health; industrialists' economic clout; workers' fear for their jobs; and the difficult issues of domestic fires and mobile railroad smoke. Many manufacturers "solved" the smoke problem by building taller smokestacks, a boon locally but a bane downwind.[71]

Industrial capitalism's peak around 1900 accompanied worries about the resource base that supported it. Britain rediscovered *The Coal Question* not because British coal ran out but because Jevons's prediction came true that America's coal production would surpass Britain's. Britain's relative decline as global leader in energy production was a sign of decline to come, as Jevons realized.

Americans reread *Man and Nature* and worried that an absolute, not relative, decline of resources threatened America's future greatness.[72] In the late 1880s, a movement gathered momentum to tackle political corruption by large corporations, labor unrest, pollution, and rapidly retreating forests. A string of Presbyterian-raised presidents—Benjamin Harrison, Grover Cleveland, Theodore Roosevelt, and Woodrow Wilson, along with Unitarian William Howard Taft—sought a government that put the common good above private enrichment.[73] Along with anti-monopoly legislation and the first national regulatory agencies and child and woman labor laws, presidents began creating forest reserves in 1891, which the Forest Service managed in the public interest after 1905. Scottish-American writer and activist John Muir popularized national parks and influenced their expansion between 1890 and 1920, overseen after 1916 by the National Park Service. At the same time, scientists began to think in terms of ecology, which is the science of reciprocal links between living things and their environment.[74] Systems thinking influenced Pinchot, who worked closely with Roosevelt to publicize a general solution to overexploitation of natural resources under the term "conservation." Roosevelt widely publicized the concept and made it central to his politics. Conservation became central to American thought about resources and increasingly to international discourse as well.[75]

Wildlife conservation rose to a crescendo in the era. Already in the 1860s and 1870s, societies for the protection of birds sprang up in Britain, France, Germany, and elsewhere, followed by the Audubon Society in the United States in 1886. Britain and Germany established nature preserves, while Roosevelt established over fifty wildlife refuges. To protect migratory birds, twelve European nations signed the International Convention for the Preservation of Useful Birds in 1902 and the United States and Canada signed the Migratory Bird Treaty in 1916. After a 1900 conference on preservation of African wildlife, the Society for Preservation of the Wild Fauna and Flora of the [British] Empire was established in 1903. A Ligue Suisse pour la Protection de la Nature was founded in 1909, which led to the foundation of the Consultative Commission for the International Protection of Nature by fourteen nations in Berne in 1914.[76] The first American state laws regulating hunting seasons and hunting and fishing licenses and catch limits were passed and multiplied in the coming decades.[77]

Conservation became an international cause in the interwar period. Local, national, and international organizations sprang up to lobby and advise. Nations approved regulations, laws, and treaties. Local, regional, and imperial wildlife conservation organizations were founded. President Franklin Roosevelt made conservation a central theme of the New Deal, embodied in the Civilian Conservation Corps. His administration pushed policies for soil conservation (especially during the great American environmental disaster known as the Dust Bowl), expansion of forest and wildlife protection, and new, wilderness-oriented national parks.[78] American political leadership of conservation faded after Roosevelt's death in 1945. The International Union for the Protection of Nature was established in 1948 in Europe, not the United States.[79]

The fading of Western industrial capitalism

Conservationists never challenged industrial capitalism itself. Life without it had become unthinkable. Conservation meant only fixing the environmental problems that industrialization had caused. Industrial capitalism was Jevons's goose laying golden eggs, even if it was a sooty, wheezing goose. Surely all the beast needed was a little fresh air, a wash, and a bit of uplifting greenspace, and it would produce eggs forever.

Industrial capitalism had bumped into limits to its growth. True, technological wonders still poured from laboratories and workshops. European aristocrats still strutted like peacocks in the unsettled years before the Great War. Nonetheless, the imperialist race to snatch up the last remaining uncolonized bits of the earth had nearly run its course, leaving only the relative boredom of governing. European and American railroad infrastructure was in place and the phenomenal growth of the American steel industry slowed. American and British coal production reached a plateau. Hugely self-confident and optimistic fifty years earlier, Western science, art, and music at century's end seemed less sure of themselves. Curious anomalies puzzled physicists. Artists abandoned an esthetics born in the early Renaissance. Germanic symphonies and operas sprawled across hours of self-indulgent self-importance.[80]

The way ahead was clouded with discontent and suppressed violence. The twentieth century rudely shouldered aside the Victorian Era and with it its hopes for an orderly and moral world. Anarchists, revolutionary socialists, and nationalists sought to advance their causes with violence and assassination. Albert Einstein's Theory of Relativity of 1905 showed how eternal external frames of reference were illusions. New art movements—Cubism, Fauvism, Expressionism—rejected conventions of nineteenth-century art. Composers abandoned traditional tonality and rhythm. Sigmund Freud shone light into the darker corners of the mind, giving lie to vaunted Western rationality.

Industrial capitalism steamed doggedly on for several more decades, through two appallingly destructive and deadly world wars and a Great Depression, during which the goose's golden eggs nearly stopped for a time. But its day was passing. After 1945, the great European empires crumbled with surprising rapidity. Steam engines vanished into museums. Smokestacks one by one stopped smoking. But, with the new century, a new version of capitalism roared to life which promoted consumption rather than production and brought new, more threatening environmental problems.

SIX

Buy Now — Pay Later

The 1920s and the rise of the automobile

In 1899, Alfred P. Sloan was the young president of Hyatt Roller Bearing Company in Newark, New Jersey, which made a unique patented bearing, when orders for wheel bearings began coming in from a new industry that was sputtering to life in the American Midwest. Automobile manufacturers wanted something better for their cars than greased wagon axles. At the time, hundreds of mechanics hand-built about 1,500 automobiles a year for an exclusive market. Difficult to start, unreliable, and poorly adapted for America's terrible rural roads, autos were an expensive toy for the rich. Nevertheless, thirty years after selling his first few dozen auto bearings, Sloan was president of General Motors as it sold 1,482,000 cars in a year.[1]

Sloan got involved in auto manufacturing just when a popular revolution in transportation was about to start. Beginning in the 1880s, rapid advances in the technology of internal combustion engines, along with steam and electric alternatives, encouraged countless mechanics and tinkerers in Germany, France, Italy, Great Britain, and the United States to mount engines on modified carriages. At the same time, bicycles and electric trolleys and streetcars had helped people to become accustomed to the idea of self-propelled vehicles. But if the car was ever to be a popular mode of transport, manufacturers needed to produce large numbers at low cost. At this, the Americans excelled. In 1895, Charles and Frank Duryea founded the Duryea Motor Wagon Company in Springfield, Massachusetts, the first company to produce identical vehicles, and sold thirteen the next year.[2] In 1902, in Lansing, Michigan, Ransom E. Olds was the first auto manufacturer to use an assembly line, mass-producing the Oldsmobile. In 1908, Thomas Edison's former chief engineer Henry Ford designed the well-built, reliable, and rugged Model T. Assembly lines produced Model Ts so cheaply that even farmers could buy them. By

1920, half the world's automobiles were Fords. Two years later, Ford sold two million Model T's. Before production ended in 1927, he had made a total of fifteen million. By 1929, 60 percent of American households owned a car, far higher than any other nation—one registered car for every five Americans.[3] Never before in history had a single product with such power to change life and landscape become ubiquitous so quickly.

To entice consumers to buy this flood of vehicles and keep buying them required creative sales techniques. The leading innovator of these techniques was not Ford, but Sloan. His focus on enticing consumers to buy put GM in the vanguard of a new stage of capitalism, consumer capitalism. Sloan also exemplified the transition from the days of the great captains of industry—or robber barons, as some styled them—often of Reformed Protestant background like Andrew Carnegie or John D. Rockefeller, to the era of the manager and corporate investor, few of whom would have a Reformed upbringing. Finally, cars embodied a revolutionary energy transition from coal to petroleum, along with a host of new environmental problems.[4]

In another sign of the future, GM was a holding company, not a manufacturer like Ford Motor Co., and its founder, William C. Durant, was a salesman, not a mechanic or tinkerer like Ford. A successful salesman of cigars and then insurance in Flint, Michigan, Durant was looking for investment opportunities when he bought a small carriage company that had designed an innovative suspension system. He pioneered a business strategy that combined high-volume, standardized products, ownership or control of key suppliers, and a nationwide network of franchised dealers. By 1906, the Durant–Dort Carriage Company was the nation's largest carriage producer. Investors asked Durant in 1904 to help a cash-strapped automaker, Buick. Durant knew little about cars but suddenly envisioned a future of carriages without horses. He took control of Buick, which soon was the largest automaker. Using Durant–Dort's business strategy, he founded General Motors in 1908 as a holding company to buy up complementary auto manufacturers and important suppliers and developed a network of dealers. Financially overextended, Durant lost control of GM when markets contracted in 1910, regained it in 1916, and then lost control again in 1920 during the postwar recession.[5]

Like Durant, Sloan was no mechanic or tinkerer. Born in 1875 in New Haven, Connecticut, son of a coffee and tea importer and grandson of a

Methodist minister, he graduated with a degree in electrical engineering from the Massachusetts Institute of Technology. A wealthy friend of his father's got him a job as a draftsman at a company he had an interest in, the Hyatt Roller Bearing Company. In 1899, Sloan's father and another investor bought Hyatt and made Sloan president. Durant bought Hyatt in 1916 and Sloan quickly become a close advisor to GM's presidents. In 1923, Sloan himself was named president of GM.

Under Sloan's guidance, GM became the world's largest industrial enterprise. In contrast to Durant the glad-handing risk-taking salesman, Sloan the humorless engineer ran GM like a well-oiled machine with many interworking parts. He coordinated the component companies in GM so that they used the same parts and body suppliers. In 1921, Sloan structured GM's auto divisions Chevrolet, Pontiac, Oldsmobile, Buick, and Cadillac to produce vehicles of ascending price and luxury. He steered overall operations but allowed men under his authority wide latitude to use their best judgment. GM's departments were coordinated but decentralized, a difficult balancing act at which Sloan excelled.

GM under Sloan transformed the auto business. In 1919, GM's chief financial officer organized the General Motors Acceptance Corporation (GMAC) to offer financing to potential customers. Sloan prompted creation of the General Motors Research Corporation (GMRC) in 1920, with engineer and inventor Charles Kettering at its head.[6] Sloan's innovations in marketing represented a change in focus from maximizing production to boosting consumption. With GMAC, consumers need not save the full price of a vehicle before buying or could buy cars more often. Sloan's idea of a range of brands from the inexpensive Chevrolet to the luxury Cadillac allowed GM to retain customers as they aged and grew wealthier. Sloan spent lavishly on advertising. By 1924, GM bought more advertising space than any other company. Sloan realized that the auto market was changing. Because the Model T's utilitarian style and color (black) barely changed from year to year, buyers could purchase a cheaper used Ford that barely differed from a new one. In 1925, GM introduced annual styling changes and choice of bright colors of GMRC-developed Duco enamel paint. Annual model changes and color options incited consumers' desire to have the latest, most stylish, most technically advanced model, which could set their cars apart from their neighbors'.

The Ford Motor Company remained more of the nineteenth century than the twentieth. Henry Ford's focus on mass production of a single high-quality item reflected the thinking of the industrial revolution. He was slow to accept the arrival of consumer capitalism. He refused on moral principles to offer financing that put people in debt or to advertise a well-built practical product whose qualities he felt spoke for themselves. The Model T lost market share to GM. When GM's sales surpassed Ford's in 1927, Ford stopped producing Model Ts, shut down the huge Rouge plant for a year to retool, and introduced the improved Model A in a range of body styles and colors. Advertising increased and in 1929 Ford began offering limited financing.[7]

Sloan steered GM to the leading edge of the transition from the industrial capitalism of James Watt and Andrew Carnegie to the consumer capitalism of Ray Kroc and Jeff Bezos, where it rode successfully for a full half a century as the world's largest industrial organization. Consumer capitalism is a supercharged version of capitalism. It is premised on selling ever more goods and services at an ever-faster pace by accelerating the speed with which money goes through consumers' hands. Born after 1870, it reached full maturity in the 1920s in the United States, where consumers spent $56.2 billion dollars in 1922 and $74.3 billion five years later.[8] Because making and using rising amounts of consumer goods required extraction of more energy and natural resources and dumping ever-greater mountains of waste, consumer capitalism forever changed humans' relationship to their planet.

A brief history of consumption

To be clear, consumer capitalism is not synonymous with consumerism, which has a long history. The first consumers were those in early agricultural societies who accumulated more wealth than needed for subsistence. Consumers acquired rare or prestige items from traders, artisans, or scribes; they built large, imposing structures; and they bought servants and services to express their glory and power, to impress rivals, and to enhance appearance. However, no one was a modern consumer yet. Aside from fine food and drink, people acquired very few things to be used up and thrown away. Occasionally, in some eras, people of merchant or upper classes kept up with the latest evanescent clothing fashions. In general, though, rarely in history have people sought items for the sake of novelty. Buyers wanted special objects, clothing, and structures to last and to be kept in their families as objects of value.

Desire for certain consumer goods laid the foundation for empires and the rise of Western capitalism. In the ancient world, a lively hemispherical trade grew up in such luxury goods as sugar, spices, Chinese silks, and Indian cottons. The fall of Rome broke those trade links with western and northern Europe, until the Crusades revived interest in exotic luxury goods. The search for those goods lured the Portuguese to Africa, India, and east Asia, where they encountered the Ming empire's vast, lively trade. Chinese silks and porcelains joined the cottons, sugar, and spices of south and southeast Asia in the holds of European ships. As the Dutch and English empires rose, Europeans learned to enjoy the novelty of unfamiliar consumer goods. Colorful Indian cottons, tobacco, tea, coffee, and chocolate slowly and then quite completely entered the lives of Europeans at all social levels. Consumerism followed trade routes. European colonists in the Americas consumed madeira and port, tea, sugar, chocolate, coffee, and European manufactured items. Africans desired European guns and Indian cotton.[9]

Women's consumption constituted a hidden motor for the eighteenth-century transition to industrial capitalism. Most products of imperialism and industrialization fed demand in the typically female domestic sphere. For the kitchen, there was tea, coffee, sugar, produce from the agricultural revolution, and china; for sewing, pins and needles; and for clothes and textiles, buttons, draperies, bedding, cotton, indigo, and other fibers and dyes. Mechanization of spinning made women's traditional textile production in the home easier until factory labor replaced it. Of course, in the eighteenth century, men, too, sought such consumer items as elegant clothing (with plenty of buttons), buckled shoes, powdered or colored wigs, and tobacco. Yet female demand lay behind a large share of the consumerism that trade and empire and manufacturing satisfied.[10]

The craving for novel foreign goods set the Industrial Revolution in motion. In his quest to imitate expensive Chinese porcelain, James Watt's and Matthew Boulton's friend Josiah Wedgwood industrialized pottery making. Restrictions on imported cotton left a consumer desire unfulfilled and helped spark mechanization of cotton production.

America led the way in moving its economic base to mass production, in creating institutions of mass consumption, and in developing methods of mass attraction and desire. In 1800, most people in Europe and the United States made their own clothes, raised their own food, and built their own homes, barns, and furniture, but a century later very few people did all of that. Most bought needed or desired items from shops. In the middle of the nineteenth century, that veritable temple to consumption, the department store, appeared in Paris and major European capitals. In the U.S., however, after the 1870s department stores sprang up in smaller towns as well as large cities. The department store was no mere overgrown country store, but a revolution in marketing. Previously, storekeepers kept items behind a counter and displayed them on request, to discourage shoplifting. Customers bought what they came to purchase. Department stores displayed items where potential buyers could see and inspect them. Attractive displays caught the eye and encouraged impulsive purchases of more items than customers had intended to buy. Because these large stores made money from volume rather than markup, prices could be cheaper than in a general store or shop. Department stores sponsored events both in and outside the store (like Macy's Thanksgiving Day parade in New York City), decorated large windows with seasonally changing displays, and designed interiors to delight and attract visitors. For American consumers on farms and small towns, Montgomery Ward in 1869 invented the mail-order catalog, a department store between covers. Joined by Sears, Roebuck, the two Chicago companies did gigantic mail-order business by the end of the century.[11]

Industrial capitalism was making way for consumer capitalism. Manufacturers had become proficient at producing merchandise cheaply and abundantly. Transportation had become so inexpensive that the outpouring of consumer goods reached to the ends of the nation or, often, of the world. So abundant were manufactured products by the late nineteenth century that industrialists confronted saturated markets. Production had overtaken consumption. Reining in production to match consumption caused unemployment, decreasing consumption, and stalled the economy. The industrial-capitalist economy contracted regularly, in 1819, 1837, 1857, 1873, 1882, 1893, and 1907. Various ways arose to encourage consumption, which culminated in the 1920s in the United States.

Figure 9. Interior view of Marshall Field & Company retail store, located at State and Washington Street, Chicago, around 1900. The department store signified the birth of consumer capitalism. Virtual palaces of consumption, they lured customers inside with elaborate window displays. Inside, consumer goods invitingly displayed enticed customers to buy, often more than they intended to. (Chicago History Museum, ICHi-039800)

GM's triumph rested on many factors, all of them hallmarks of consumer capitalism, including financialization, consumer debt, transition to petroleum, corporate research, planned obsolescence, and advertising. As a holding company, GM participated in the trend toward financialization of capitalism. In financialization, corporations and investors see greater profits in financial actions, such as buying and selling companies, than in investing directly in trade or manufacturing. At the top of corporate hierarchies, managers like Sloan grew more common and more powerful and inventors, entrepreneurs, mechanics, or tinkerers rarer. Consumer capitalism's eternal cycle of increasing consumption depended only partly on steadily rising incomes that it produced. Debt put more money into the hands of buyers, and GMAC was just one of many methods of selling things by putting customers into debt which sprang up in the 1920s. At the same time, the United States began its second great energy transition, from coal to petroleum, which the auto industry pushed more than any other. Only petroleum could provide the immense amounts of energy that the endless cycle of production and consumption demanded. For its part, the GMRC foreshadowed a new emphasis on corporate-funded research and development of novel substances and machinery, formalizing and accelerating the activities formerly done mainly by garage tinkerers and kitchen inventors. Moreover, GM's promotion of new purchases by making older models seem obsolete or unfashionable mirrored increasing use of planned obsolescence to sell new products—consumer capitalism's endless cycle of buying and throwing away. Finally, GM's gigantic expenditures in advertisements reflected the vital importance of advertising to the functioning and growth of consumer capitalism. As if to drive home the message, the Great Depression of 1929 showed how, when consumer spending slowed, consumer capitalism collapsed like a deflated balloon.

Concentrating and coordinating consumer capitalism

Financialization emerged to promote efficiency and rationalization of industries. Billy Durant put General Motors together with the same methods used by financiers like J. P. Morgan that reshaped the global capitalist economy in the period between 1850 and the 1930s. In the late nineteenth century, overproduction threatened profits. To regulate competition and guarantee profits, investors bought up competing companies. Corporate consolidation crested in the United States in the late 1890s, where, often through the trust, a legal and financial instrument pioneered by John D. Rockefeller's Standard Oil, a wave of mergers brought several thousand companies under the control of fewer than two hundred. J. P. Morgan's creation of the holding company U.S. Steel created the largest corporation in the world and the first billion-dollar corporation. Morgan bought out Andrew Carnegie in 1901, enabling Carnegie to retire and give away his huge fortune to philanthropic causes, and combined Carnegie Steel with several competitors, with the goal of rationalizing the industry. Between 1880 and 1920, income from finance increased from two percent to four percent of gross domestic product. Durant observed the strategies of financiers like Morgan and began buying and selling companies on the stock market. Buying Buick set him on the road to founding General Motors.[12]

Consolidation and corporate expansion like those that Morgan and Durant engineered exploded in the 1920s in the United States, when income from finance jumped from four to six percent of GDP. Large corporate enterprises even began to take over Main Street. Chain stores proliferated in all kinds of businesses. Chain stores allowed economy of scale and attracted customer with nationwide advertising. People bought clothes or shoes in chain department stores, got their drugs in chain drugstores, watched the latest movies in chain theater, ate at chain restaurants, and stayed in chain hotels. Tens of thousands of A&P and Kroger grocery stores spread across the nation. Montgomery Ward, Sears, and J. C. Penney opened hundreds of stores to feed and complement their mail-order businesses. Large department stores built branches in nearby communities, even as the Federated Department Stores company bought up many local department stores. To fill the shelves of all these stores, investment bankers like Lehman Brothers and Goldman Sachs facilitated or pushed mergers of small makers of consumer products into such giants as Colgate–Palmolive–Peet, General Mills, and Borden's.[13]

The financial sector laid the foundations for financialization and consolidation. Stock exchanges, like those founded in Amsterdam, London, and New York in the seventeenth, eighteenth, and early nineteenth centuries, had given businesses easier access to capital, created a market in stock shares, and separated ownership from management. Privately funded for-profit incorporated commercial banks appeared early in the nineteenth century across western Europe and provided another source of capital. They multiplied particularly rapidly in the U.S.[14] As the reliability and speed of communication improved with cheap postal rates, telegraphs, transoceanic cables, telephones, and wireless communication, global financial centers emerged in London and Amsterdam and after 1870 in New York.[15] Capital moved with increasing ease around the world, seeking the best returns on investment. Financialization can be hugely profitable for investors yet often does nothing to develop the economy or promote innovative company policies. Thus it was that huge sums of English capital went abroad even as English industrial innovation lagged behind Germany and America.[16]

Funding desire with consumer debt

GM created GMAC because it realized that people could not become customers until they had money in their pockets to spend. To get more money to potential consumers faster, people would have to have greater access to cash than what was in their pockets, purses, or bank accounts. In the 1920s, easy credit drew almost everyone into the orbit of consumer capitalism.

American consumers' unprecedented borrowing spree of the 1920s depended on the necessary institutions that bankers and investors on both sides of the Atlantic had been setting up. Without them, only the most prosperous classes could borrow. A century earlier, lenders began to risk loans to the less wealthy. Lending institutions grew more numerous and began to offer consumer bank services. As insurance companies grew common, by mid-century life insurance became a popular way to save. Various forerunners of savings and loans banks and mutual savings banks appeared, where workers could keep savings and borrow for such large purchases as houses. Merchants, middlemen, and peddlers also extended credit (often at usurious rates) to those who never set foot inside a bank or were too poor for banks to lend to.

Manufacturers and retailers thought up new ways to finance purchases of durable goods, beginning in the United States, where people tended to go into debt much more often than in Europe. Installment plans for furniture appeared by 1807. By the 1850s, people could buy farm equipment and pianos on installments. When in 1856 the Singer Sewing Machine Company offered installment plans, sales tripled.[17] After 1900, the number of retailers and mail-order companies that sold goods on installment plans increased rapidly.[18] Around the same time, sales finance companies sprang up to provide installment loans for large consumer goods, which a decade later they extended to automobiles.[19] Mass production of cars required mass sales, which banks would not finance, so the Maxwell Motor Company first offered installment plans in 1916. The largest and most important finance company, GMAC, founded in 1919, within a decade ran most other sales finance companies out of business.[20] In 1920, about one-quarter of Americans had bought something on installment plans, commonly refrigerators, appliances, radios, phonographs, clothes, and automobiles. A decade later, Americans bought $7 billion worth of consumer goods on time.[21] Most who bought on installment plans, with their unfavorable interest rates, tended to be poorer and least able to take on debt, but the plans made it possible to own radios or clothes otherwise beyond their means.[22]

The 1920s birthed a new way to entice people to consume by going into debt: charge plates, the ancestors of credit cards. In the late nineteenth century, department stores offered credit to reliable customers. At the turn of the century, to speed up transactions, they issued round metal numbered tokens to customers, which, however, might be stolen or used by unauthorized persons. In 1928, Filene's Department Store in Boston issued to its customers small charge plates embossed with identifying information. Other stores followed suit, with each charge plate valid only for the merchant that issued them. Some banks experimented with bank cards in the 1930s but could not overcome problems of fraud prevention, rapid payment to merchants, tracking purchases of large numbers of customers who could essentially get loans at will, and making a profit on cards. After World War II, these hurdles were finally overcome and the first universal credit card appeared, Diners Club.[23]

Powering consumption: energy revolutions

Petroleum fed the global rise and triumph of consumer capitalism. The U.S. in the 1920s and Europe after World War II transitioned from coal to petroleum. Petroleum replaced coal in large part because it is much simpler to produce, process, move, and sell. After a well is drilled, oil flows on its own or is pumped with virtually no labor, quite in contrast with coal. Oil needs little labor to store and transport, especially by pipelines, again unlike bulky, heavy coal, so troublesome to transport, store, deliver to customers, and use. Coal also is an impractical fuel to power many forms of transportation. Finally, burning gasoline produces no ash, soot, or grime. Oil fields required no large workforce that might threaten to strike and bring the economy to a standstill.[24] Heating oil freed homeowners and maintenance men from the work of shoveling coal into furnaces. This transition from coal to oil helped produce the unprecedented economic booms of the 1920s and again after 1945. Coal prices could not be severed from the cost of labor and rose with rising wages. Wage gains had less impact on the relatively stable price of oil. The growing gap between rising wages and stable energy prices gave workers more disposable income to spend on consumption.[25]

The age of petroleum began with kerosene, which chemists learned to extract from oil in the 1850s. Kerosene replaced expensive whale oil and greatly lightened hunting pressure on those huge beasts. Edwin Drake successfully drilled the first modern well at Oil Creek near Titusville, Pennsylvania, on August 27, 1859, to supply a Pittsburgh refinery making kerosene.[26] Local Scotch-Irish Presbyterians drilled or invested in thousands of wells in the region. One of these investors was Andrew Carnegie, who invested his profits in iron and steel.[27]

Rockefeller saw he could make steadier profits in refining oil than in the risky boom-and-bust cycles of drilling and production. He began buying refineries in Cleveland, Ohio, the closest port to Pennsylvania oil fields. Hoping to make the oil industry more rational, efficient, and profitable, Rockefeller perfected the legal instrument of the trust to buy up competitors and integrate them into his Standard Oil Company. Oilmen he drove out of business fanned out across the nation and developed competing fields in California, Texas, Oklahoma, and elsewhere. The glut of oil slaked the nation's growing thirst for petroleum products.[28] Global oil production, too, expanded, as fields were discovered or exploited in Russia, Galicia, Canada, Dutch East Indies (Indonesia), Persia (Iran), Mexico, and Venezuela.

Nothing served as both symbol and catalyst of the great twentieth-century energy transition from coal to oil like the automobile. The cars that Ford, GM, and other automakers produced in such immense quantities moved with the potent energy from a new form of fuel: gasoline. Refiners distilled gasoline from petroleum, of which it makes up a small fraction. Until development of the internal combustion engine, perilously flammable gasoline had so few practical uses that refiners often poured it into rivers at night. The rapid growth of the auto industry put pressure on limited gasoline supplies. Engineers at Standard of Indiana, one of the companies created by the antitrust breakup of Standard Oil in 1911, figured out how to "crack" petroleum and more than double the yield of gasoline from a barrel of oil. Huge new oil-field discoveries in

Texas and the American west averted a fuel crisis.[29] Without the quickly expanding oil industry that pumped, transported, refined, and sold gasoline at service stations in every town and crossroads across the nation, the age of the auto of the 1920s could never have happened.

Electricity was another revolution in the making. Electricity is a bit of a genie. It has countless advantages over other forms of energy, except for its limited portability. Practically any form of energy can produce it—organic, fossil, or nuclear fuel, solar heat or light, or motion of air or water. Clean, unpolluting, silent, and safer than alternatives, it can be stored in batteries or transmitted instantaneously from one place to another. High-energy lines can simply and easily be stepped down to low power uses. It can power motors of virtually any size and can produce light, heat, and sound. In circuits it allows unfathomably complex logical operations in computers and other devices. Finally, electricity is instantly available through the flick of a switch.

Electric power only slowly become a consumer desire. Businesses used it first. Electrification liberated machinery from dependence on belts connected to overhead shafts. In the metals industry, electricity electroplated one metal on another or purified metals like aluminum. Electric motors ran streetcars and elevators, making skyscrapers possible. Scottish-American Alexander Graham Bell's telephone of 1876 and Edison's light bulb of 1878 multiplied in businesses and cities. Energy historian Vaclav Smil describes the swiftness that Edison acted to put his invention into use:

> A durable light bulb was a mere beginning: in the three years after its unveiling Edison filed nearly 90 new patents for filaments and lamps, 60 dealing with magneto- or dynamo-electric machines, 14 for the system of lighting, 12 for electricity distribution, and 10 for . . . electric meters and motors. The first electricity-generating plant, built by Edison's London company at Holborn Viaduct, started to transmit power on January 12, 1882. New York's Pearl Street Station, commissioned on September 4 of the same year, was the first American thermal power plant. A month after its opening it energized some 1,300 light bulbs in the city's financial district, and a year later more than 11,000 lights were wired.[30]

Cities around the world soon electrified and bought thousands of bulbs. From urban electrical networks grew national and international grids.

Around 1920, Americans suddenly regarded light bulbs as essential for homes, not just businesses and streetlighting. Farmers languished in the dark until the New Deal's Rural Electrification Agency in 1935 brought lights and electricity even to them.[31] People now had become consumers of electricity. In the 1920s in the United States, electric consumer goods rapidly multiplied. (European consumers lagged considerably. Americans generated over half the world's electricity by 1929.)[32] Washers, refrigerators, radios, phonographs, stoves and ranges, irons, and vacuum cleaners appeared rarely in the homes of 1914 and commonly in middle-class urban homes of 1930.[33]

Petroleum and electric power enabled regions, such as California, which lacked cheaply accessible coal to bypass the coal era altogether. With abundant petroleum reserves and hydropower from the Sierra Nevada, California sped almost directly past coal from organic energy to oil, gas, and electricity.[34]

Plastic fantastic consumer capitalism

Petroleum yielded an even greater store of substances not found in nature than coal tar had. Researchers at corporate research laboratories like GMRC discovered this new world of artificial materials. In the early twentieth century, chemists learned to crack petroleum's long chain molecules into shorter ones. Then after World War I, the United States seized and sold the assets and patents of formerly dominant German chemical companies and dominated the global chemical industry thereafter. In the 1920s, American corporations sent a flood of products made from these artificial compounds into the market. America's vast oil and gas reserves gave it a natural incentive to discover and sell these substances.[35]

The first common petroleum-based plastic was Bakelite, an invention of Belgian-American chemist Leo Baekeland, who began manufacturing it in 1909. Bakelite molded to nearly any shape, resisted water and heat, and was durable. In the 1920s, it could be found in telephone bodies, in radio cases, dials, and knobs, in electrical insulators and panels, in silverware handles, in toiletry sets. Automobiles had Bakelite in ignitions, spark plugs, paint, battery terminals, ashtrays, and steering wheels. American production of Bakelite and other synthetic resins rose from 1.6 million pounds in 1921 to 33 million pounds in 1929. By 1940, acrylic, melamine, vinyl, nylon, and neoprene had joined the plastics parade.[36]

In the 1920s, GM chemists invented other compounds that in time would have significant environmental impact. In 1921, GMRC engineer Thomas Midgley, Jr., found that leaded gasoline solved the problem of engine knock, which decreased efficiency and damaged engines. Midgley's other signal achievement was development in 1928 of the chlorofluorocarbon Freon for refrigerators in GM's Frigidaire division. Freon replaced the toxic, flammable, and explosive chemicals then in use. GM contracted with Du Pont to manufacture both chemicals. Despite its experience with dangerous products, Du Pont at first struggled to safely produce the deadly poisonous leaded gasoline, which poisoned and killed five Du Pont and two GMRC employees before the company learned sufficiently safe production procedures.[37]

Until the 1920s, common household items were made from something found in nature. Countless consumer products contained animal products (wool, silk, leather, feathers, horn, tortoise shells, whalebone, or ivory) or plant products (wood, cotton, or linen). Bakelite and the plastics that proliferated in its wake freed consumer capitalism from the limits of nature. The natural world could never have supplied enough raw materials to feed consumer capitalism's need for perpetual growth. Unfortunately, these synthetic products do not rot or break down but persist in the environment for decades or even millennia. The chemicals that created them were toxic, carcinogenic, or harmful. Exhaust from leaded gasoline was a nerve toxin and Freon destroyed beneficial stratospheric ozone. Artificial consumer products hid a host of problems for humans and the natural world.

How to get people to buy what they already own

Consumer capitalism depends on constant purchases of consumer goods. Yet a consumer only needs one automobile, radio, stove, oven, refrigerator, vacuum cleaner, iron, or any of many durable goods. Even with easy money, market saturation loomed early in the 1920s. Sloan again saw a solution. GM's Chevrolet had failed to compete against Ford's Model T, so in 1923 designers and engineers redesigned the look of the car and fixed several mechanical flaws. Now more stylish than the utilitarian, only-in-black Model T, the car sold. Sloan noticed. A car was the most expensive consumer good most people owned, and they appreciated that could signal success, prestige, and status. Soon GM offered all its cars in a variety in bright colors and with styles and features that changed yearly. Formerly, only such technical advances as Charles Kettering's electric starter had made car models obsolete. With annual style changes, GM had hit upon an effective incentive to get consumers to replace vehicles with many years of life ahead of them. Other manufacturers adopted GM's strategy. Henry Ford dropped the Model T for the more stylish and colorful Model A and acquiesced in 1933 to annual style changes. The idea caught on elsewhere. Products from washers and refrigerators to radios and lamps soon appeared in the popular streamlined style of Art Deco.[38]

Getting consumers to replace goods that inherently had no style required the development of products that failed. Here Edison's General Electric Company was the leader. Founded in 1889, GE faced stiff competition after Edison's light-bulb patent expired in 1894. GE controlled or bought up strategic patents to keep American light-bulb makers at bay. To fend off antitrust actions, GE improved bulb technology and lowered prices, so that, in the 1920s, the cheap light bulb for the first time truly became a product for mass consumption. GE promoted its bulbs aggressively in the 1920s, but once all houses were brightly lit, owners had no need to buy more. In 1924, GE took advantage of the postwar chaos and mistrust among European lighting companies to create Phoebus S.A. of Switzerland, a cartel that determined quotas, prices, and quality. Phoebus set the standard life of a bulb at 1,000 hours, a reduction of about twenty percent, which required light-bulb owners to replace bulbs more often. Planned obsolescence had arrived.[39]

Planned obsolescence became a common tactic during the Depression. Most Americans and Europeans wanted to keep their level of consumption the same or higher as before. With less income, they had to buy cheaper goods.[40] Manufacturers often substituted cheaper materials in production to make products cheaper. Cheaper materials hastened the day when consumers would have to replace products. What began as a strategy to boost the economy became a permanent practice.[41]

Even more profitable for a manufacturer would be something that was obsolete as soon as it was used—a disposable product. Such items appeared in the 1920s as well. The first was the safety razor with a blade that was discarded rather than stropped or honed for reuse. King Camp Gillette invented a razor with a disposable blade that attracted little attention until he won a contract to supply razors and blades to the army in World War I. Since soldiers could keep their razors after the war, Gillette suddenly had thousands of customers regularly throwing their used blades away. Soon Gillette was selling hundreds of millions of blades around the world.[42]

The war also brought disposable products to women. Kimberly-Clark had developed an absorbent celluloid product for gas-mask filters and bandages and had a large stock when the war ended. In 1920, the company marketed the material as Kotex, the first inexpensive disposable sanitary napkin for women. Four years later, the company developed a disposable handkerchief of the same material, which it called Kleenex, promoted for removing makeup. Noticing that women used Kleenex for blowing their noses, Kimberly-Clark marketed it as a disposable handkerchief. Other manufacturers followed with their own disposable products. Johnson and Johnson introduced Band-Aid in 1924. Modess brought out a more comfortable sanitary napkin in 1927 and, in 1934, Tampax marketed the first tampon.

The throwaway society had been born.

Creating desire: advertising, consumer capitalism's propaganda

"Advertising is one of the vital organs of our entire economic and social system. It certainly is the vocal organ by which industry sings its songs of beguilement," said President (and former Secretary of Commerce) Herbert Hoover to a 1930 national meeting of advertisers.

> The purpose of advertising is to create desire, and from the torments of desire there at once emerges additional demand and from demand you pull upon increasing production and distribution. By the stimulants of advertising which you administer you have stirred the lethargy of the old law of supply and demand until you have transformed cottage industries into mass production. From enlarged diffusion of articles and services you cheapen costs and thereby you are a part of the dynamic force which creates higher standards of living.[43]

Hoover put his finger on an essential component of consumer capitalism. For all that corporate consolidation, easy credit, an energy revolution, artificial materials, and planned obsolescence did to increase consumer spending and accelerate the economy, modern consumer capitalism could not operate without a powerful propaganda agent—advertising—to weaken cultural and moral barriers to spending on unnecessary or self-indulgent items. Advertising flourished with the dramatic expansion in the 1920s of mass media: movies, mass magazines, music and records, and radio. Mass media and advertising did more than promote products. In the interwar era they painted a beguiling image of a consumer society, often called the "American Way of Life" or the "American Dream."[44]

Advertising was by no means new in the 1920s, but the transition from industrial to consumer capitalism transformed it in that decade. Modern media and advertising had grown up with the British industrial revolution. American advertising's contribution was to advance beyond simply announcing a product's availability to encouraging consumption.[45] Modern advertising agencies appeared in the U.S. in 1877. Print advertisements began to supplement text with illustrations. The development of paper bags and cardboard cartons allowed packaging of many items previously sold out of barrels or bins, giving consumers consistent quality and clean products. Bags and boxes invited manufacturers to print distinctive text and images on them that promoted both the product and provided a public image for a faceless corporation. After 1900, American advertising grew more sophisticated as advances in printing and photography allowed clearer and more elaborate illustrations. Psychological and market research began to affect advertising copy.

The U.S. government enlisted advertising agencies to support the Great War, making connection between advertising and propaganda explicit. So effective were the methods they developed that they shaped postwar advertising, but also propaganda supporting the Red Scare of 1919 and Joseph Goebbels's Nazi propaganda. Sometimes ads combined commercial and political propaganda, as when Sloan's 1930s GM ads included anti–New Deal messages.[46]

After 1920, advertising grew more pervasive. American expenditure on advertising as a percentage of gross domestic product exceeded any other nation. Normally constituting between two and 2.5 percent of U.S. peacetime GDP, advertising expenditures reached a historic high of three

Figure 10. Would your husband marry you again? 1921. This advertisement for Palmolive soap in *Harper's Bazaar* used novel methods of salesmanship. Soaps vary little in their formulas, so manufacturers use marketing techniques to create an image for their product. This ad plays upon women's insecurity. The image of a well-dressed couple suggests to the reader that wealthy, sophisticated women use Palmolive. A sultry, topless, and very white "Egyptian" beauty in the inset draws attention to the palm and olive oils in the soap. (Duke University Libraries)

percent of GDP in the 1920s.[47] Newspaper advertising revenue more than quadrupled between 1914 and 1929. This flood of money into advertising gave owners of mass media huge amounts of capital to expand and reach everyone in society but the poorest and most remote. The purpose of media changed from informing, entertaining, or persuading to selling. Media was growing inescapable.[48]

Advertising also became more persuasive. Ads exploited insecurities and vanity for such personal products as makeup, soap, mouthwash, deodorant, and shampoo. They incited the desire to look more prosperous or sophisticated, showing images of men or women in formal evening attire examining refrigerators or vacuum cleaners or cars. Advertisements could promote the wish to feel or appear more up-to-date and modern with the latest styles and designs or introducing new products. They fed a vague aspiration for comfort and convenience, notably for throwaway products. A successful advertising campaign made something that no one had thought they needed into a necessary possession, whether a second car, an extension telephone, or shampoo.[49]

Corporate commercial propaganda encountered serious cultural resistance. American advertising agencies found that techniques developed for the United States did not always work as well abroad. European and Australian advertisers admired American methods but, under the conviction that they would be less effective in the culture of their own countries, delayed adopting them for another generation.[50] Then, too, as the European economy did not recover from the war as quickly as the American and showed signs of weakness even before 1929, people had less expendable income. Finally, the United States had a huge market compared to the smaller, fragmented national markets of Europe, which made mass media and mass advertisements much easier and more profitable.

Consumer capitalism did make inroads into European empires. American movies, for example, found audiences around the world. However, cultural and racial attitudes affected imperial policy in such a way as to leave the colonized peoples without the resources to purchase many consumer items. Had attitudes and policies been different, the colonies might have been a much larger market for the products of mother countries. In Africa, south and east Asia, Russia, and Latin America, the foreignness of consumerism made it both attractive and strange, at least until someone found a way to make it locally culturally appropriate.[51]

Even in the United States, serious cultural resistance to corporate propaganda developed. Advertising competed with family, religion, schools, and peers to propagate consumer values. American Protestant and republican ideals had long promoted the values of self-denial, industry, sobriety, thrift, and community and condemned selfishness, prideful self-regard, waste, avarice, and greed. Such values do not sell consumer products. Advertising and mass entertainment promote pleasure, self-gratification, entertainment, consumption, and the individual.[52] Thorsten Veblen, Sinclair Lewis, Vance Packard, David Riesman, John Kenneth Galbraith, and other commentators and intellectuals loudly decried the direction of modern American consumer society.

Ironically, American Protestantism in many ways paved the way for modern consumer propaganda and conceivably helped make the United States the bastion of consumer capitalism that it has long been. In the early nineteenth century, a wave of religious revivalism swept the country. Revival preachers focused intently on inciting emotional religious experiences in individuals, who by those means attained salvation. Evangelicalism's profoundly individualistic mission lacked the communal parish basis of the Protestant denominations that had dominated colonial religious life. To reach and save sinners, revivalists used the latest in communication technology and techniques. Religious presses used the latest technology to flood the country with religious propaganda of tracts, pamphlets, devotional books, magazines, and newspapers, each with text and illustrations to persuade readers of their truth. In the twentieth century, evangelists quickly seized upon such new technology as radios and then television to spread their message. Advertisers really just followed the path that revivalists blazed.[53]

Landscapes of consumer capitalism

Consumer capitalism rapidly magnified human environmental impacts. To begin with, automobiles transformed life and landscape like no other consumer good. They drove animals and people from roadways and streets. Today, cities often restrict autos in favor of walking, biking, and public transportation, but a century ago, people welcomed cars.

Unlike horses, cars did not foul city streets with thirty to fifty pounds of manure each every day, leave waste that bred disease-spreading flies, need care and feeding when not in use, die on the streets, or panic and bolt uncontrollably down the street.[54] Horses and wagons disappeared from cities and farms and in the United States were mostly gone by the 1940s. Pedestrians had to go, too. Fast, maneuverable, and quiet on their rubber tires, cars and trucks killed unwary people by the hundreds, especially children. No one had ever needed to learn to look before stepping off curbs to cross streets. Crowds of pedestrians crossed anywhere they liked, dodging slow-moving horse-drawn vehicles as they went. Under pressure from automobile clubs, towns and cities passed traffic laws that ensured that people yielded to the right of cars to have the streets, rather than the other way round.

Also, since American politicians won elections by keeping fares low, chronically underfunded municipal public transit was crowded and poorly maintained. Americans abandoned streetcars and urban trains. As towns, cities, and suburbs expanded, planners designed them to ease automobile traffic. Suburban streets were widened. Houses acquired garages. No longer tied to streetcar lines or train stations, towns spread out in a pattern soon dubbed "urban sprawl." Shopping and industry joined the exodus from the city. Developments of detached single-family dwellings sprang up around major American cities. Already in 1922, 135,000 houses in suburbs of 65 cities depended on cars to commute to work.[55] White upper and middle classes escaped the crowds and pollution of cities by moving to outer districts with cleaner air and water, a trend that accelerated after the war.[56] Restricted-access highways enticed more people out of public transportation into their cars. Los Angeles grew almost entirely by virtue of such developments.

The American countryside changed as well. Car owners wished to drive out into the country, where awful American roads thwarted many a Sunday pleasure drive. To keep taxes low, local authorities made roads only as good as necessary for farmers' horses and wagons. Drivers and auto clubs demanded better roads. In the 1920s, car lobbyists got Congress to federally fund and design a nationwide system of paved highways. The highway system essentially subsidized the trucking industry, whose heavy vehicles had been tearing up most roads. Trucks competed with railroads, whose policies and prices were overseen by the Interstate Commerce

Commission, which lacked regulatory authority over trucking. Railroads lost freight business to trucks, which had greater flexibility and speed and could deliver goods from door to door, although were relatively inefficient in long hauls. Railroads continued a relative decline until deregulation in 1980.

Northern and, to a far lesser degree, southern European cities sprawled almost as much as American cities. Between 1921 and 1931, for example, London's population grew from 7.5 to 8.2 million but the urban area doubled, as people with more modest incomes followed the wealthier classes who earlier had fled the congested city center to the suburbs. The phrase "urban sprawl" was coined for London in 1934.[57]

Denser populations in Europe and Japan encouraged alternative transportation to automobiles. A GM study team concluded in 1929 that the German auto market was eighteen years behind America's. Narrow, winding, crowded city streets discouraged automobile traffic, and ownership of autos per capita remained significantly lower than in the United States. Bicycles remained popular. Also, military leaders pushed nations like Germany and France to invest in efficient railroad networks for wartime movement of troops and supplies, which benefited railroad passengers. On the other hand, in the 1930s, Nazi Germany designed and built the autobahn road system designed for autos only and gave Germans the opportunity to open the throttle or take leisure drives. Town planners in Britain and Japan also began moving toward better road networks. Still, widespread auto ownership generally awaited the return of prosperity after World War II.[58]

Mechanization of farming dramatically changed rural landscapes, too. When small multipurpose internal-combustion tractors appeared during the Great War, American farmers enthusiastically adopted them everywhere except for the poverty-stricken Southern states, which relied on mules until after World War II. The number of horses on farms fell by over half between 1918 and 1940. In 1915, farmers devoted about 93 million acres to feeding animals, about a quarter of farm acreage. By 1960, when the transition from animal to mechanical power was complete, only four million acres supported animals. Farmers could devote more land to crops, a portion of whose value went toward purchase of the tractor and of external inputs like fuel, lubricants, spare parts, and tires. Farm production rose and food prices fell. Generators and the arrival of electricity with the New Deal after 1935 encouraged purchase of electric labor-saving machinery, especially milking machines, which 12,000 farmers owned in 1910 and 175,000 in 1940.[59]

Declining numbers of farm animals deprived farmers of use of their manure for fertilization, which went hand-in-hand with other changes. In 1911, German chemists Fritz Haber and Carl Bosch succeeded in the first industrial-scale synthesis of ammonia, a source of nitrogen. Slowly, fertilizer products came onto the market after the war, but American farmers balked at the cost and did not widely adopt nitrogen fertilizers until after 1945. In the meantime, to fertilize their fields, farmers relied on another external input, superphosphates derived from bones, less effective but one-third the cost.[60] Chemical companies and the Chemical War Service experimented with poison gas as an insecticide, which failed with a couple of exceptions. The tear gas chloropicrin was marketed as a fumigant and paradichlorobenzene gained acceptance against peach-tree borers on farms and against clothes moths in homes. Gas delivery by airplane also proved successful for crop dusting. In the late 1930s, a chemist in the Swiss chemical company Geigy (later Ciba-Geigy) discovered the insecticide dichlorodiphenyltrichloroethane (DDT) and another chemist in the German company I. G. Farben discovered organophosphates, a class that included such pesticides as parathion and malathion, as well as nerve gases such as sarin. These pesticides found widespread application after the war.[61]

Sating consumer capitalism's appetite for natural resources

Easy money, planned obsolescence, and propaganda solved the problem of mass overproduction at a profit and made consumer capitalism run. Money moved more quickly through the economy. When consumers bought something, money passed into the pockets of store clerks, factory workers, managers, owners, investors, and almost everyone else, who now had money for consumption. The genius of consumer capitalism was to speed the cycle up. Consumption yielded higher employment, wages, and standards of living.

The natural environment paid a steep price for prosperity. Speeding up the economy also raised the rate of consumption of natural resources. Production of consumer goods required energy and raw materials. After purchase, durable consumer goods from autos to electric fans continued to consume raw material and energy. Rapidly rising production demanded great quantities of natural resources.

The two biggest boom industries of the Roaring Twenties, automobiles and construction, both required steel, which demanded more iron ore, coal, and coke. The building boom included construction of the world's three tallest structures, the Manhattan Trust Building, the Chrysler Building, and the Empire State Building, all completed between 1930 and 1931. The Empire State Building alone required 58,000 tons of steel. The air of steel-producing cities was black with prosperity.[62]

Automobiles also demanded rubber for tires, hoses, wires, and gaskets. In the early 1920s, Americans owned 85 percent of the world's automobiles and used 75 percent of the world's rubber, primarily from British, Dutch, and French colonies in southeast Asia. The British Empire produced 77 percent of the global rubber supply. After Britain restricted rubber exports to support its price in 1922, American tire and auto manufacturers sought to produce rubber under their own control. In 1926, Henry Ford established the rubber plantation Fordlandia in the Amazon rainforest in Brazil, a costly disaster because of labor problems, tropical diseases, and a native fungus that attacked rubber monocultures. With greater success, Harvey Firestone of Firestone Tire and Rubber developed a huge rubber plantation in Liberia, a nation founded by freed American slaves.[63]

Prosperity demanded oceans of petroleum. Every motor vehicle needed petroleum products to operate—engine oil, grease, and fuel. The automobile was a godsend to the oil industry just at a time when electric illumination was diminishing demand for kerosene. Just before the Great War, the world's navies switched from coal to fuel oil for practical reasons and greater economy east of Suez. Internal combustion engines took to the air in 1903, when the American Wright brothers flew the first successful powered airplane. Aircraft design leapt rapidly forward. In the decade after the end of World War I, aircraft crossed the Atlantic, Pacific, and all the inhabited continents. Commercial air routes crisscrossed the globe by 1925.

The United States had abundant supplies of petroleum in many huge oil fields. In Europe, Baku and Romania had oil fields but with current technology no exploitable oil fields lay under most of the rest of Europe.

Oil in the Dutch East Indies made Royal Dutch Shell a major global oil company. British Petroleum controlled oil in Persia. Despite these fields, and others in Venezuela, Mexico, and the Arabian peninsula, in 1929 the United States produced seven out of ten of the world's barrels of oil and almost all its natural gas.[64]

While petroleum has many advantages over coal, coal is safer to handle, does not create major environmental problems when it spills, and rarely explodes (except for gas or coal dust in mines). From the beginning, petroleum proved itself dangerous and messy to extract and transport. Every oil discovery encouraged a rush of drillers eager to get in early and get rich. Drilling and production were usually hasty and careless and sometimes undercapitalized. Vegetation, soil, structures, and men around early oil fields were black with petroleum, and streams glistened with an oily rainbow sheen. In new fields, oil could blow out drilling equipment and "gush" uncontrollably for days or weeks. Natural gas in a gusher could explode in a shower of burning oil droplets, turning people into human torches and igniting wooden derricks and buildings. Kerosene lamps, wood stoves, and smoking threatened to ignite fumes and gas. Oil stored in ponds seeped into groundwater and poisoned the soil. Oil spilled when workers transferred it to and from wagons, railroad tank cars, ships and boats, and refineries, and again in transmission to customers. Pipelines leaked. Finally, customers had to dispose of old oil and used petroleum products, which usually went into sewers and waterways.[65] Petroleum and its products contain many toxic and carcinogenic chemicals, at first poorly understood. Refinery workers and townspeople suffered from elevated cancer rates. Adding lead to gasoline raised an outcry from doctors, which fell on deaf ears for decades. Finally, tailpipe emissions did not attract much concern until the end of World War II.

The other great new power source of the age, electricity, is much cleaner to transfer and use, but often not to produce. Utilities powered electric grids with coal and, later, natural gas, which added to the environmental problems of air pollution and coal ash disposal. Where coal was scarce and topography allowed, as in California and Japan, dams drove generators. America generated a third of the world's hydro-electric power in 1929.[66] In the 1920s and 1930s the United States began a dam-building spree in western states that lasted into the 1970s and inspired countless imitators around the globe. Beginning with Hoover Dam, the federal government began building immense dams and by the 1930s was constructing the world's five largest dams simultaneously. Along with tremendous electric power, dams provided reliable water supplies for cities, irrigation, and navigation. In the Tennessee Valley, they also served as erosion control. Pollution-free dams caused many other environmental problems. They drowned towns and farms in settled states, disrupted stream flows, changed water temperatures, diminished or killed off native species of fish and shellfish, submerged wetland habitats, and blocked anadromous fish like salmon. In the arid west, evaporation lost much valuable water and concentrated minerals and salt.[67]

The problem of waste

The gargantuan input of natural resources into consumer capitalism's mass production of consumer goods matched a monstrous output of waste on the other end. Consumers bought new goods and threw out used, outmoded, unwanted, and disposable items, which had to go somewhere. Smoking chimneys from factories working to keep up with demand filled the air with choking black clouds. Drainpipes poured industrial and household wastes into rivers, streams, and harbors.

The stream of trash and refuse grew rapidly. American production of solid waste per person rose fifteen percent between 1920 and 1940. As consumers moved out of compact city centers into sprawling suburbs, whose boundaries hindered centralized policies, collection of trash grew more difficult and expensive. Haulers had to travel farther to dump their loads. As open wagons and trucks hauled away refuse, winds scattered trash and dust. Identifying a place to dump refuse posed practical and political problems. Towns with greater wealth and political power sited dumps among poorer, weaker populations. Open dumps were cheap but attracted vermin, stank, contaminated groundwater, and sometimes caught fire. In the 1920s, some authorities began the practice of burying mixed waste. Modern sanitary landfills, called "controlled tipping," appeared first in Great Britain in the 1920s. In the United States, a few cities adopted them in the 1930s. In 1935, General Electric introduced an electric garbage disposer for home sinks, the Disposall, which sent kitchen waste into streams and rivers.[68] Many towns and suburbs built incinerators, a solution that grew more popular by the late 1930s. Supposedly modern, sanitary, and relatively free of pollution, incinerators appeared in Great Britain, Germany, France, and the U.S.[69]

Hazardous industrial water pollution drew increasing attention in the interwar years. A report in 1923 identified 248 water supplies that industrial waste had contaminated. Production of new chemicals from petroleum sent new wastes into streams. The mix of wastes in the water supply daunted investigators, who found them difficult to analyze or assess for danger to public health. Consequently, solutions to the problem of industrial water pollution awaited the era after World War II. By contrast, knowledge of the connection between human waste and disease prompted increasing number of cities and towns to build sewage treatment plants in both Great Britain and the United States.[70]

When consumer capitalism stops

American consumer capitalism had fueled the world's greatest economic boom. Unfortunately, the stock-market crash of 1929 so alarmed consumers that many decided to wait to make any major purchases. Spending on durable goods paused. Inventories grew. Factory orders fell. Workers lost jobs. Consumer spending dropped. Banks failed. Unpaid consumer debt led to repossessions of cars and durable goods and foreclosures of real estate that no longer had buyers.[71] A self-reinforcing downward spiral began that lasted four years, until much of the American economy had ground nearly to a halt. Consumers spent only a quarter as much on furniture and a fifth as much on radios and musical instruments as they had at the height of 1920s prosperity. Steel production in 1932 was twelve percent of capacity and auto production in 1933 was 35 percent of 1929 levels.[72] The crashing American economy pulled many of the props out from under the weak European economy, which collapsed in some nations and slowed in all.

Consumer capitalism resembles a sort of vast Ponzi scheme. It requires constant growth to survive. If consumption ever stops or slows, the global economic system stumbles and totters. After only a decade of unprecedented growth, the Roaring Twenties whimpered into the Great Depression. As industrial production declined to a small fraction of capacity, smokestacks stopped smoking and skies cleared. Nature got a decade's respite from consumer capitalism's insatiable hunger for natural resources. The another war even more destructive and all-encompassing than the last got the global economy roaring again. The return of factory production in World War II brought back regular paychecks. Smoke-filled air seemed a reasonable price to pay for economic security.

The modern combination of mass production and mass consumption is called Fordism, but it might better be named Sloanism, after the true author of the system. Sloan's management brought GM to the top of auto production in the 1920s and 1930s. A supporter of consumerist individualism and suspicious of demands of the community, he gave money to right-wing opposition to the New Deal. Sloan managed to deflect postwar questions about his handling of GM's German interests under the Nazis, which was either overly sanguine or disturbingly cooperative. He remained at GM's helm as prosperity returned after the war, retired in 1956, and died in 1966. By then, however, GM was growing conservative and ossified. When consumer capitalism charged back in the United States in the 1940s and 1950s and in Europe in the 1950s and 1960s, a rising generation raised in prosperity would decry the plundered, polluted planet. GM would struggle to meet the challenges of a changing world.

SEVEN

Stepping on the Gas

Consumer capitalism gets into everything

Los Angeles, California, grew up with consumer capitalism. Founded by the Spanish in 1781, L.A. was tiny until the arrival of the railroad and the discovery of oil in the late nineteenth century. Land was cheap and flat, the weather was almost always pleasant, and natural amenities like endless beaches and the San Bernadino Mountains beckoned. A 1913 aqueduct that stole water from farmers in the Owens Valley east of the Sierra gave the city enough water to grow. The Hollywood movie industry widely publicized L.A. in the 1920s and 1930s. During the Depression, thousands of mostly conservative evangelical southerners came looking for work.[1] Thanks to a defense industry and military bases to support war in the Pacific and then the Cold War, Los Angeles mushroomed during and after World War II. Many former servicemen who had passed through or been stationed there or in San Diego just to the south decided to stay. In the postwar era, the city attracted millions more. Between 1945 and 1970, L.A. epitomized postwar consumer capitalism, its non–Reformed Protestant architects, and its environmental impacts more than any other place.

L.A.'s rise paralleled the expansion of the automobile industry. The end of streetcar service after World War II made car ownership nearly imperative. New services like drive-in movies, churches, and banks popped up that did not require leaving a vehicle. Drive-in restaurants had appeared in the 1930s, where carhops took orders from people in their cars and brought them their food. The modern fast-food industry was born in 1948 in suburban Anaheim. Brothers Maurice and Richard McDonald, sons of Irish immigrants, opened a drive-in restaurant in 1948 with no carhops, a cheap but limited menu (hamburgers a mere 15¢), and simplified, unskilled kitchen work. The McDonalds served food to be eaten with fingers and served on disposable paper plates or in paper cups, and could dispense with utensils and crockery, which could break or be stolen. Business boomed. They began to sell franchises.

A milkshake-machine salesman, Ray Kroc, son of Czech immigrants, was impressed by the number of machines he sold to the McDonalds. In 1954 he bought the rights to franchise their restaurant. Like William Durant an aggressive and successful salesman, Kroc sold thousands of McDonald's franchises around the country over the next two decades. Determined to run the business according to his own ideas, he bought the brothers out in 1961. Kroc focused on marketing to children, knowing that they would bring adults with them. Ads ran during children's television programs. A clown, Ronald McDonald, became the company mascot in 1965. A small playground designed by a former designer for Walt Disney was attached to new restaurants. As fast-food restaurant chains saturated the U.S., McDonald's sought new markets overseas. Kroc retired in 1974 but his restaurants have continued to multiply until they sell in well over a hundred countries, on every inhabited continent.

The McDonald's chain's tremendous success inspired imitators of its fast-food model almost immediately. Burger King, Wendy's, Kentucky Fried Chicken, Dunkin' Donuts, Carl's Jr., Taco Bell, and many, many others followed, and entrepreneurs from many industries copied Kroc's franchise system. Advertisements for franchised businesses grew ubiquitous on television, in print, and, eventually, on the Internet. As freeway systems grew, franchises sprang up along roads and highways, then invaded city centers. Independent businesses in cities and towns lost business or shut their doors under the onslaught of their advertising and reliably uniform products.[2]

By 1970, consumer capitalism had captured almost everything in American society and was on the march out of the U.S. into the wider world. From fast food to packed processed food on supermarket shelves, food had become a packaged, advertised, cheap, quick, and easily purchased item, like soap or deodorant was back in the 1920s. Environmental impacts were hidden but huge. Fast-food franchises each demanded great quantities of hamburgers, buns, french fries, condiments, milkshakes, or whatever food the franchise specialized in, and all of it of reliably consistent quality, taste, and price. Farmers had to produce animals and grains to suit corporate bulk-buyers, which meant mass production of meat using hormones, antibiotics, and grain-fed feedlots and of grain with pesticides and herbicides and identical strains of wheat. Farms and ranches grew larger as commodity prices dropped. And so it went throughout the American economy. The 1960s would be a self-consciously consumerist age like none before.

Postwar consumer capitalism before 1970

Burgeoning consumer capitalism made the world that Kroc thrived in. World War II had boosted the American economy and morale. In Japan, East Asia, the Soviet Union, and Europe, however, shell-shocked inhabitants faced the decades-long task of rebuilding their war-wasted cities and economies, after which consumer capitalism would conquer them, too. The Cold-War-inspired American aid program known as the Marshall Plan sent billions of dollars of grants and loans to bolster the economies of western Europe and East Asia. More importantly, however, beginning with the New Deal policies of the Franklin Roosevelt administration in the United States and after 1945 in all the world's democracies, governments regulated markets and large businesses and instituted systems of social welfare. These policies kept streams of cash flowing into consumers' pockets.

The non-Communist Allies made crucial agreements to promote free trade, in the belief that protectionism and economic collapse had fueled the rise of fascism and caused World War II. The Bretton Woods monetary agreement of 1944 created the World Bank and the International Monetary Fund to promote development in poorer countries. The General Agreement on Tariffs and Trade of 1947 built the international frameworks within which the global capitalist economy has grown ever since. So began stable economic growth that lasted until the 1970s without major economic downturns.

The world was changing quickly. Europe stood diminished next to the great American colossus. It had taken European nations five centuries to build their globe-girdling empires. They collapsed almost completely within five decades. Two horrendously bloody and destructive wars, Japanese eviction of Europeans from their Asian possessions, and Europe's weakened postwar economies damaged respect for European power. Across the global South, new nations full of hopeful optimism sought models for developing their own economies and institutions.

While some of them suffered from chaos and corruption, others have remained stable and prosperous. Most postcolonial economies continue to rely on extractive industries (mining, plantations, timber) often owned by foreign companies. Economic development and participation in global consumer capitalism has risen steadily if unevenly.

Anxiety over the Cold War and the nuclear arms race shadowed the era, but overall an optimistic mood prevailed. Social change was in the air, along with a sense of possibility and hope that a better world was at hand. Optimism and faith in the future found visible expression in a worldwide Baby Boom, the sudden rapid rise in fertility rates that lasted from the mid-1940s to the mid-1960s.[3] Youthful, energetic President John F. Kennedy seemed to embody the hopeful idealistic dynamism of the early 1960s, when unprecedented American power and prosperity made anything seem within its grasp. In the United States, racial equality made the greatest advances in a century. The Civil Rights movement inspired such notable events as the 1963 March on Washington and the 1964 Freedom Summer and such landmark legislation as the Civil Rights Act of 1964 and the Voting Rights Act of 1965. The nation's history of racial injustice seemed finally to be in its last chapters. Lyndon Johnson succeeded Kennedy in 1963. In 1964 he proclaimed the goal of a Great Society from which racism, poverty, ignorance, and pollution would finally be banished. Dramatic, historic legislation ensued that addressed all those problems.

The buoyant mood spread beyond the industrialized democracies. In the Soviet Union, after Stalin's death in 1953, Nikita Khrushchev released many political prisoners. The Soviet economy grew rapidly and availability of consumer items increased, albeit of low quality and volume. Although Khrushchev lost power in 1964, de-Stalinization spread elsewhere in Warsaw Pact countries. Communist Czechoslovakia gradually loosened restrictions and control, which led to the liberal reforms of Alexander Dubček and cultural efflorescence of the Prague Spring of 1968. In China, after the disaster of Mao Zedong's Great Leap Forward, Mao sought to regain power against rising opposing factions in the Community Party by starting the Cultural Revolution. He tapped into the idealism of the young to root out supposed bourgeois pro-capitalist elements in the party who threatened the revolution.

This era also saw the peak of Euro-American confidence in Enlightenment logic and the superiority of its own institutions. Certainly,

Western science and technology dazzled the globe with their successes and achievements, while leaders of poor countries everywhere envied the power and prosperity that industrialization had brought. As the Cold War intensified, the United States and Soviet Union saw themselves in an existential contest for global influence. American culture, especially in music (rock and roll and rhythm and blues, especially) and entertainment (e.g., Mickey Mouse), seeped into cultures almost everywhere, even through the Iron Curtain into Communist eastern Europe and the Soviet Union. However, the putative scientific logic of Marxist socialism held an undeniable attraction to intellectuals everywhere. To counter, the U.S. emphasized Communism's hostility to religion and tendency to authoritarianism, and held out constitutional democracy, human rights, and free-market capitalism as principles of the "Free World." American foreign policy was thoroughly inconsistent here, since presidents were often willing to sacrifice the first two if it would preserve the last. Nevertheless, both the free and Communist nations expressed great confidence in the power of rational systems.

The New Deal and World War II had opened America's economic throttle to maximum, so that, when the war ended, consumer capitalism flew thundering down the track like a drag racer on high octane fuel. The New Deal had many important economic effects. New programs and government support of labor unions led wages and productivity to rise and working hours to fall. The Federal government invested massively in infrastructure, including roads, bridges, government buildings, huge hydroelectric dams, rural electrification, conservation, and flood control. Progressive taxation and social programs like Social Security reduced income inequality. Over the next thirty years, American personal incomes at all levels rose at approximately the same rates without significant growth of income inequality. Finally, new agencies and programs lowered economic risks for everyone from investors to farmers to homeowners. The GI Bill financed a major jump in education advancement, as well as in home ownership and other benefits. Moreover, wartime Federal government investment paid for so much new productive capacity between 1940 and 1945 that capital stock increased by half (the number of machine tools alone doubled), all of which was newer, more modern, and more productive than prewar private capital stock. Additionally, the demands of war production had forced companies to learn to operate more efficiently. In the century from 1870 to 1970, the United States came to dominate the globe in technology and innovation. The flight of scientific, technical, and cultural talent from Europe during the 1930s and 1940s benefited the U.S. more than any other nation, so that, when peace returned, the nation stood as the (almost) unchallenged center of scientific and technological innovation.[4]

The war boosted the chemical companies. Peacetime found them with new plants making novel chemicals, and they sought civilian markets for them. New artificial plastics and fibers poured forth from corporate research laboratories to be developed into consumer products. The miracle substance of the age, plastic found uses in all kinds of products. Chemical companies also developed and marketed a variety of new, powerful insecticides to farmers and the public. Corporate chemists created novel chemicals by the hundreds and thousands, which companies marketed aggressively for dyes and a myriad other manufacturing and consumer uses.

Electronic consumer technology also moved rapidly ahead. Soon after their invention in the late 1940s, transistors replaced vacuum tubes in televisions, radios, and other consumer electronics. Now smaller and more affordable, these devices proliferated in daily life. IBM's room-sized computers, developed from wartime progenitors, filled climate-controlled basements in banks and universities, while small computers accompanied astronauts into space in the 1960s. ARPANET (Advanced Research Projects Agency Network of the U.S. Department of Defense), the ancestor of the Internet, took shape in the late 1960s and hosted the first e-mail in 1971. Medicine also made rapid strides in surgical techniques (such as organ transplants and open-heart surgery), antibiotics, vaccines (against polio, encephalitis, influenza, measles, mumps, and rubella), radiotherapy and radioisotope therapy, chemotherapy, and hormonal contraception (the Pill). With generous government and military investment, aerospace technology made such rapid strides that, by 1970, hundreds of huge Boeing 747s flew around the world and Apollo spacecraft were taking men to the moon and back. Artificial satellites revolutionized astronomy, meteorology, communication, and earth reconnaissance.[5]

Despite the innovations and inventions, the postwar American boom looked much like the boom of the 1920s. Millions upon millions of consumers bought new (or newly affordable) consumer products, from televisions to clothes-dryers to air conditioners. Newly prosperous young people bought cars and homes. Suburbs sprawled across farmland around all major cities. Cities and states built roads for suburbanites to move around and commute to jobs in city centers. The federal government invested heavily in road construction, including the extensive interstate highway system after 1956. Local governments and railroads eliminated most public transportation, except busses. Outside the centers of older cities, daily life depended on cars. Car-oriented businesses multiplied, from gas stations to fast-food franchises like McDonald's. Shopping centers and large stores with plenty of parking sprang up. Cities lost tax base as the middle class headed for the suburbs. Businesses deserted city centers and opened in commercial strips along major roadways in the suburbs or in office parks surrounded by a sea of parking spaces. For GM, this was its golden age.

There was a worm in the apple of postwar prosperity, slowly growing out of sight until it appeared in triumph in the late 1970s. The regulations and government activism of the New Deal (not to mention the apparent success of the Soviet experiment in the early 1930s) so alarmed certain wealthy corporate leaders, Alfred Sloan among them, that they began to develop a propaganda network to promote weak government and low taxes.[6] Corporate ads promoted individualism and equated consumption with democratic freedoms.[7] Millionaires in industries that feared regulations' threat to profits, such as the oil and gas and the tobacco industries, poured money into fundamentalist and evangelical denominations, which began a postwar rise to national cultural and political influence. These churches promoted individual holiness rather than social improvement.[8] Over the decades, millionaires and corporations set up an increasingly dense and widespread system of foundations, institutes, think-tanks, and news organizations to influence public opinion, legislatures, politicians, and the courts. The Cold War, portrayed as a contest between godly freedom against godless Communism, allowed businessmen to wrap capitalism in a flag and give it a cross to hold high.[9] Their greatest triumph still lay in the future, after this era of American growth.

Propaganda for consumption and capitalism

The sophistication and power of postwar salesmanship advanced in great strides. To make purchases easier than ever for consumers without cash, Diners Club introduced the first general credit card in 1950. Many similar cards followed from banks, oil companies, and other corporations. Shopping malls sprang up across the land, especially in new, sprawling suburbs, to simplify shopping for car-dependent consumers. In the 1960s, shoppers could buy in air-conditioned comfort in enclosed malls whose design allowed merchants to subtly influence shoppers and encourage unplanned purchases.

Corporate commercial and political propaganda entered a new era of power and influence with the rapid spread of television in American households at the expense of radio, movies, and print media. American television networks were private and depended wholly on advertising revenue. While entertainment held viewers' attention, advertisers attempted to sell them products. Advertising agencies still used older methods of selling while they exploited possibilities of images and short storylines.[10] The power of the new medium showed itself in a famous cigarette advertising campaign. Lagging behind other large cigarette companies, in 1954 Philip Morris began a television advertising campaign for the Marlboro brand, which had a mere one percent share of the market. The campaign emphasized masculinity and sales rose precipitously. A decade later, the Leo Burnett ad agency struck gold when it switched the ads to scenes of grittily realistic cowboys on the 6666 Ranch in west Texas. Viewers watched smoking cowboys ride horses, herd cattle, rest by a fire, and fish to the theme music from the movie *The Magnificent Seven.* Adjusted to local cultures elsewhere, this advertising campaign made Marlboro the world's best-selling cigarette brand—all without one word about the product's qualities. "Come to Marlboro Country," the ads said, and millions of smokers came.[11] Despite this demonstration of its powers of persuasion, television advertising could not perform miracles. No amount of advertising could convince consumers to buy the notorious Ford Edsel. (Perhaps they should have shown a smoking cowboy driving one.)

Nevertheless, so effective was television advertising that politicians, political parties, and government propagandists borrowed its methods. Politicians and political operatives discovered the power of television in

Figure 11. A family gathers around a television set in the 1950s. By 1960, rare was the American home without a television, perhaps the most powerful propaganda device for consumer capitalism ever invented. Like these children, most of the Western postwar generation grew up in the cool glow of its cathode ray tube. (H. Armstrong Roberts. Alamy, CMRToY)

1952, when the first political conventions were televised. They realized that events that built excitement on convention floors often did not play well on television and staged future conventions more for television cameras than for delegates. In that year, also, Richard Nixon rescued his political career with his televised "Checkers Speech" to defend himself against corruption allegations. A real moment of truth was the first presidential debate, on September 26, 1960, between candidates Richard Nixon and John F. Kennedy on live television, which many believed made Kennedy president. Lyndon Johnson lacked Kennedy's charisma and relied heavily on negative advertising to win the White House in 1964. As people often noted in the 1960s, campaigns sold politicians to the public like brands of soap. In 1968, Nixon ran again for president. As chronicled in the aptly titled bestseller *The Selling of the President 1968* by Joe McGinnis, Nixon hired Roger Ailes to craft his media image. So well did Ailes succeed that he would be hired by Ronald Reagan, George H. W. Bush, and the infant cable channel Fox News. Ailes became a powerful conservative media consultant and master right-wing propagandist.[12]

Television was also emblematic of a new kind of corporation, the service corporation, that would surge to great size and influence after 1970. Broadcasting corporations offered programming as a medium for advertisements. Newspapers and radio networks had prepared the way as purveyors of information, entertainment, and political persuasion in order to sell advertisements. None commanded the power over the public that moving images on a television tube wielded, more like cinema than text. And TV was free to the viewer until the rise of cable television in the 1980s. Advertisements moved further from offering informative descriptions of products to selling status, sex, and image.

The postwar generation in developed countries around the world was the first raised in front of the television screen. More thoroughly than any previous generation, the young deeply absorbed the values of consumer capitalism. Even when they believed they were dropping out from consumer culture in the 1960s and 1970s, they promoted its values: individualism in lifestyle, individual freedom to "find themselves" or "actualize their potential," and self-gratification in drugs and sex. The counterculture was easy to commercialize and quickly entered consumer culture. It shaped style, entertainment, and much else. In the next century, television would yield primacy for advertising and consumerist-political propaganda to a new medium, the Internet. The Internet would be as freely available as early television, on a device small enough to be carried in pockets and purses, never separated from its owner—advertisers' ultimate dream.

World energy transitions

Petroleum, the liquid that makes consumer capitalism run, now displaced coal worldwide. Western Europe and Japan followed the early start of the United States. Oil's proportion of Europe's energy use rose from 23 percent in 1955 to 60 percent in 1972, while in Japan it leapt from seven percent in 1950 to 70 percent twenty years later. Global demand for petroleum in 1972 was over five and a half times as great as in 1949. With plenty of coal but no oil (yet), Britain reluctantly replaced coal for most of its energy needs because oil was cheaper and, in a confirmation of Jevons's argument, using costlier coal would yield competitive advantage to foreign industries.[13]

Discoveries of huge new oil fields more than kept up with demand. Geologists located giant new oil fields in Ghawar, Saudi Arabia, in 1948, Daqing, China, in 1959, West Siberia in 1960, Prudhoe Bay, Alaska, U.S., in 1968 and the North Sea in 1969.[14] These joined countless smaller new fields all over the world. Production in the non-Communist world increased from 8.7 million barrels a day in 1948 to 42 million barrels a day in 1972. American production continued to expand, but large discoveries abroad pushed down its share of world production from 64 percent to 22 percent in the same period.[15]

The gargantuan oil fields of the Middle East were practically on western Europe's doorstep, which allowed Europeans, who in 1948 got 77 percent of their oil from the Western hemisphere, to buy 80 percent from the Middle East a few years later.[16] However, there was a problem. The rise of Arab nationalism worried Europe. Control of the Suez Canal, through which two-thirds of Europe's oil passed, was vital. Egypt seized the canal in 1956 to pay for the Aswan Dam that Western nations refused to fund. Syria, to make a political point, shut the main pipeline to Europe for 24 hours.[17] Concerned about their vulnerable energy supply, Europe turned to supertankers, which Japanese shipmakers had recently developed, and which could profitably and efficiently take oil around Africa to Europe. The supertanker, a major advance in transportation technology, has been vital for bringing cheap petroleum products to the world.[18]

Globalization of consumer capitalism

Another development in transportation facilitated consumer capitalism's growing triumph: the container ship. After World War II, American railroads had begun "piggybacking" truck trailers on flatcars to take back some of the freight cargo lost to trucking. In 1955, Malcolm McLean of McLean Trucking saw possibilities in shipping trailers by sea as well and bought control of a shipping company. Rather than rolling trailers on and off his ships, however, he developed the idea of detaching the wheels and running gear and hoisting the wheelless trailers, or containers, aboard ships by a crane. McLean's Land–Sea Service sailed the first container ship in 1956. Not only were container ships extremely efficient, but they made loading and unloading cargo far faster and cheaper. McLean's engineers redesigned the containers so that they could bear the weight of containers stacked on them and latch securely onto each other, features that also allowed gantry cranes to easily latch onto and hoist them during loading and unloading. In 1963, McLean released the patent for the container design to promote standardization of the industry.[19]

The container ship transformed international trade. Japan's isolation no longer prevented it from becoming a major exporter. South Korea, Taiwan, Malaysia, Singapore, and the Philippines soon followed, with China joining the group at century's end.[20] To bring down shipping costs, ships grew larger until by the early 2000s individual ships carried over 15,000 containers.

Consumer capitalism did not come to the Soviet Union and its eastern European allies, but consumerism did. Back in the 1930s, authorities awarded consumer goods to productive workers, political leaders, and other exemplars of socialist values. After World War II, rising incomes stoked desire to consume. Authorities promised stores stocked with goods, automobiles, and other consumer items, and used the consumer-capitalist West as the standard that they promised their people they would overtake. Radio and television programs and smuggled goods also helped create an unfulfilled envy of Western prosperity and goods. Socialist authorities were unable to provide the goods they promised, partly because Communism's political base was heavy industry. Consumer goods remained in short supply and of poor quality.[21]

Beyond North America, Europe, and Japan, consumer capitalism has not advanced uniformly. In the three decades after independence in 1948, consumer capitalism arrived in India with development of urban middle and upper classes. After India opened its autarkic economy in 1991, electricity and consumer goods gradually reached rural villages. While still very piecemeal, consumerism has completely transformed traditional Indian life in a mere generation or two. India's huge neighbor to the north, China, has not really joined the consumer capitalist world yet, even with its unique government-directed hybrid Communist capitalism. It of course now produces many of the world's consumer goods. Yet when it shifted the cost of once-free housing, education, and healthcare onto households after 1997, policies discouraged frivolous Western-style self-gratification through buying. In Latin America and Africa, consumer capitalism has progressed steadily, if slowly and, again, unevenly.[22]

Prosperity's environmental costs: oil

As consumer capitalism gathered speed after 1945, so did environmental problems. All the ecological horrors that shadowed the good times of the 1920s—air and water pollution, chemical use, plastics and other synthetic materials, urban sprawl, overflowing landfills, dams and irrigation—and entirely new ones like radioactive substances loomed larger and more formidable than ever. Consumer capitalist society increasingly choked on the waste products of its own golden success.

Only after 1945 did the full costs of the energy transition from coal to oil reveal themselves. Ironically, many of the very qualities that made gas and oil so attractive produced as many problems as they solved. As liquids, petroleum and its products are hard to contain. Toxic and inflammable, oil leaks from wellheads, barrels, pipelines, and tanks. It spills when transferred between containers and vehicles. Refineries release toxic and carcinogenic vapors into nearby neighborhoods and spill or dump noxious and dangerous chemicals into waterways. Companies and scientists knew of these problems early on and in the 1920s and 1930s worked with limited success to address them.[23]

On March 18, 1967, the world discovered another price of dependence on oil. The *Torrey Canyon*, a supertanker in top condition, in fine weather and broad daylight, under the command of a well-experienced captain, carrying almost 120,000 tons of Kuwait crude to a refinery in England, struck the Seven Stones Reef, a well-known hazard. Thanks to winds that blew 85 percent of the oil into open ocean, only a small portion reached the Cornish and Breton coasts. That was enough to blacken a hundred and forty miles of coast and beaches. The ship was privately owned and in international waters. The ship's owner, Union Oil of California, and operator, British Petroleum, hoped to recover the ship or at least not endanger insurance coverage, but salvage operations failed when the ship broke up. No authority had prepared for this day, so response was confused and delayed. Local, national, military, and civilian officials lacked any coordinated plan, dithered over uncertainty about who would pay costs, and worried about political repercussions. Finally, the government decided to

Figure 12. *The Torrey Canyon,* full of crude oil from the Middle East, breaks up on a reef off Cornwall, England, in 1967. It was the first accident for the new, large-capacity oil tanker called the supertanker. The environmental disaster it unleashed alerted the world to the environmental price it would have to pay for consumer capitalism's thirst for energy. (Alamy, B4TJFJ)

use bombs, napalm, and burning kerosene to destroy the tanker and burn off the remaining oil. The *Torrey Canyon* was built to be fireproof, so they only succeeded in burning about 40,000 tons. Volunteers, government employees, and soldiers swarmed the rocks and beaches. They treated the slick and blackened beaches with 800,000 gallons of a detergent used for small harbor spills, but which was toxic to marine life. Over 30,000 seabirds (guillemots, razorbills, greater auks, and others) and many marine animals died. When the oil reached Britanny, the high mortality that dispersants and detergents caused in England led the French to a much more cautious, conservative, effective, and far less deadly approach.[24] Although many more spills would follow in succeeding decades, some of them dwarfing it in size, the *Torrey Canyon* spill shocked because it was the first.

Less than two years later, a second shock rocked the United States and the world. On January 28, 1969, oil workers employed by the unlucky Union Oil Company were drilling into the seabed from Platform A five miles off the coast of Santa Barbara, California. They had just removed between 500 and 700 feet of piping from the hole they were drilling. At the end of the hole, 3,479 feet below, artesian pressure blew out drilling mud and blowout prevention equipment and spewed dangerously inflammable natural gas and oil into the air and sea. Drillers dropped the pipe back into the hole and crimped it to plug the hole. However, gas and oil forced their way around the pipe and up through fissures in the ocean bedrock and bubbled to the surface uncontrollably. Every day for ten days, tens and perhaps hundreds of thousands of gallons of oil flowed into the sea. Although local residents had long worried about possible oil spills, government and oil company officials had not planned for them. Too few workers and too little equipment were on hand to respond. Union Oil officials delayed notifying the government so that the company could make its own decisions. Their first plan was to contain the oil so that it did not reach the coast and more or less hope for a miracle. The miracle did not occur. Santa Barbara was a small city with a university and a wealthy, white, Republican population, upwind from Los Angeles smog and with no polluting industries of its own. It occupies a spectacularly beautiful site between the coast range and the sea. Directly in front of the eyes and cameras of a politically influential populace, oil covered rocks, beaches, thousands of seabirds, seals, and whales. Workers managed to cap the well, but oil continued to seep into the sea for many months afterward. Although the oil industry argued that the spill had caused little biological damage, many scientists and the photos in the press argued to the contrary.[25]

Automobiles encouraged urban sprawl, which not only spawned a plague of franchises like McDonald's but had many environmental impacts of its own. Postwar sprawl was first a problem in America, where prosperity returned quickly, a baby boom was underway, and New Deal programs made home ownership much easier. Elsewhere, in Europe and East Asia, authorities were rebuilding destroyed cities, which allowed much greater central control over the dimensions and direction of urban growth. Also, few cities there grew much in the years after the war, in contrast with rapidly growing American urban areas. Los Angeles, for example, grew from four million in 1940 to eight million in 1970. Growth was particularly remarkable (and uncontrolled) in the cities of the southernmost American states, or Sunbelt, which air conditioning had made livable and Cold-War military spending more prosperous.[26]

Developers built suburbs quickly and cheaply. Bulldozers cleared away trees and leveled tracts to prepare for rapid assembly-line construction of roads, houses, shopping malls, and parking lots. Roads, parking lots, and buildings covered the ground with surfaces impervious to rainfall. Rain could not soak slowly into the ground and eroded soil or ran off so quickly that floods swamped low-lying communities. Unknowing homebuyers bought houses atop filled-in watercourses and suffered from drainage problems. Housing subdivisions replaced meadows, fields, groves, and forests, the very natural amenities that had made the area attractive in the first place. Because suburban growth outran city services, sewage in many suburbs went into individual septic tanks. Septic tanks could overflow or back up into homes. Where houses had not been connected to city water systems, foaming detergents or polluted water often seeped from septic tanks and contaminated household wells. Dispersed populations also use much more energy than concentrated ones, especially for transportation and for heating and cooling.[27]

Another problem first appeared in July 1943 in Los Angeles. The air turned brownish and stung the eyes. L.A.'s low-density housing had encouraged construction of the world's best roads and freeways as well as widespread ownership of cars. Residents had moved to Los Angeles in large part because it enjoyed a mild, sunny climate, clean air, natural beauty, and amenities like beaches and mountains. After the smog incident, civic and business leaders and real estate developers founded the Citizen Smog Advisory Committee to defend a city whose reputation for healthy air was essential to its growth and prosperity. Under its influence, California passed the Air Pollution Control Act in 1947, which authorized L.A. to establish the Los Angeles Air Pollution Control District, the first in the U.S. Suspicion that wartime industry caused the smog led to controls that did nothing to reduce the problem. In 1950, a scientist at the California Institute of Technology determined that automobiles and oil refineries produced chemicals that sunlight and moisture turned into smog. The oil and gas industry replied with vigorous denials. Only in 1956 did research confirm definitively that cars and refineries were the culprits. Los Angeles and the state of California became national and world leaders in the battle against smog (but not against cars or freeways).[28]

The world's newly sprawling cities all soon were gasping in clouds of smog of their own. City dwellers looked up into skies that petroleum had cleared of coal smoke but that vehicles now dimmed with smog. The easiest solution, a return to public transportation, was out of reach. No one wanted to give up cars, and the auto and oil industries fought any potential regulation of their products. Car manufacturers instead sought technological fixes, such as the catalytic converter, developed in 1954, which removed offending chemicals from exhaust. However, leaded gasoline fouled the converter. GM owned 50 percent of Ethyl Corporation and profited handsomely from leaded gasoline. So, until the 1970s, the auto industry obstructed even a technological solution to smog in the U.S. Abroad, many more decades passed before leaded gas disappeared from gas stations.[29]

Ubiquitous burning of petroleum products exacerbated a very different problem that scientists now could measure. American science was flush with cash in the 1950s and 1960s, thanks to the Cold War. Some of this money trickled into what seemed at the time like a marginal scientific enterprise, that of investigating a theory of the atmosphere. Nineteenth-century scientists had discovered that carbon dioxide and water vapor were opaque to heat and like a blanket blocked enough of the sun's warmth from radiating back into space to moderate the earth's temperature. It seemed likely that burning fossil fuels would increase the concentration of carbon dioxide enough to affect this process, but no one could figure out how to get accurate readings of its levels in the constantly moving and changing atmosphere to prove the idea. Roger Revelle of the Scripps Oceanographic Institute in La Jolla, near San Diego, California, coordinated efforts of various geophysicists to think about the problem. In 1957, one of them, Charles David Keeling, set up a sensitive device on the top of Hawaii's Mauna Loa, 11,000 feet above the sea, and precisely, accurately, and reliably measured carbon dioxide without the transient changes in concentration that bedeviled measurements at lower elevations. By 1960, Keeling reported that carbon dioxide was indeed increasing year by year. The measurements continue today and confirm that, as humans burn exponentially growing quantities of fossil fuels, carbon dioxide is warming the planet.[30]

Prosperity's environmental costs: land

Chain restaurants like McDonald's, with their cheap, mass-produced food, represented only one way that consumer capitalism changed the growing, selling, and eating of food and degraded land and soil. In the twenty-five years after 1950, world population grew from about 2.5 billion to four billion people.[31] Food production kept abreast of the growth and kept more people better fed than ever. In the U.S., the average share of income spent on food had declined from around seventeen percent in 1960 to less than ten percent in 2022.[32] How was this accomplished? Since most of the world's best arable land was already in production in 1950, people had to get more food out of essentially the same resources. The changes in agricultural practices that spread around the globe became known in 1968 as the Green Revolution.

The Green Revolution was born in the United States. Already in 1940, American farmers were the world's most productive. Between 1950 and 1970, the American agricultural workforce fell by half but yields of almost every major farm product doubled—wheat, corn (maize), cotton, and milk. Over the previous century, the country had built a network of agricultural schools and colleges, experimental and research stations, and extension agents which produced and promulgated improved varieties, all under the watchful and approving eye of the Department of Agriculture. Mechanization had advanced steadily since the nineteenth century until, after World War II, mule- and horse-drawn plows cut their last furrows. Chemical companies that had built up production capacity for the military now turned their salesmen loose on farmers to sell them the latest fertilizers, pesticides, fungicides, and herbicides. New agricultural chemicals proliferated abundantly.[33]

The Americanization of world agriculture began when President Franklin Roosevelt sent Vice-President-elect Henry A. Wallace to represent the U.S. at the 1940 inauguration of Mexican President Manuel Ávila Camacho. Camacho asked Wallace, a former Secretary of Agriculture, for assistance in improving Mexican agriculture and ending rural poverty. Wallace facilitated the Rockefeller Foundation's supervision of an agricultural improvement program by Mexican agricultural scientists, which after various problems Americans took over in the late 1940s. One of them, researcher Norman Borlaug, led the effort to find a high-yield variety of wheat that grew well in all regions of Mexico. He chose wheat, not the corn that most Mexican subsistence farmers raised, because the Mexican government wanted large wheat farms to produce wheat for export, not local consumption. Borlaug's wheat variety suited large farms with access to credit, machinery, and irrigation, to grow wheat in extensive monocultures with chemical fertilizers and pesticides. Other nations, especially India, expressed interest in Borlaug's wheat, and the Green Revolution was off and running. Wheat yields around the world rose quickly. High-yielding varieties of rice and corn soon followed in east and southeast Asia.[34]

Environmental impacts of this high-yield food production system were many. The farmers of both the Green Revolution and American agriculture applied copious amounts of chemical fertilizer and pesticides, because high-yield crops exhausted soils and large commercial fields attracted pests. To maximize yields, farmers often overapplied nitrogen fertilizer, and the excess washed into watercourses and groundwater. In streams, lakes, and bays, it fed harmful blooms of algae, while nitrates in groundwater endangered the health of those who consume it. Pesticides endangered the health and lives of farmworkers. They also killed pests and beneficial insects without distinction. When birds and fish ate insects killed or poisoned by pesticides, the chemicals worked their way up the food chain, accumulating at higher concentrations as they progress up to apex predators. Like fertilizers, pesticides got into runoff when it rains, poisoned aquatic food chains, and contaminated drinking water. Irrigation depleted groundwater, causing subsidence. The hot sun concentrated minerals in irrigation water and over time turned soil saline and sterile. Finally, careless or uncontrolled chemical use left residues on or in fruits and vegetables, which consumers ate along with their produce.[35]

Consumer capitalism also transformed food and eating, especially in the hurried and mobile society of the United States. (In much of Europe, in contrast, local food traditions long kept a much stronger hold on food consumption.) Common ownership of refrigerators made daily trips to the grocer for the day's meals unnecessary. Chain grocery stores displaced local grocers and filled shelves with canned, packaged, frozen, dried, or freeze-dried products, often preserved with techniques developed for the military during World War II. Frozen "TV dinners" in aluminum trays and colorful packaging promised ease and convenience, at the expense of flavor. When consumers grew concerned about the damage to their health from the high loads of salt, fat, and sugar that companies added in convenience foods to compensate for blandness, food corporations marketed low-salt, low-calorie, and low-fat alternatives for which factory-made chemicals now provided flavor.

The packaging of such processed foods doubled as advertising space. Large food corporations sold high-profit mass-produced packaged foods like chips and soda, whose popularity spread in large part because a large proportion of the price went to promotional advertising. Promotional advertising also drove expansion of McDonald's and other fast-food franchises, with their cheap and convenient fare in disposable wrappings and containers. Corporations began selling disposable packaging for home kitchens, previously only used for store packaging. Advertisements instructed housewives how to use disposable parchment paper, wax paper, aluminum foil, plastic wrap, and wax-paper or plastic food bags, along with disposable paper towels and napkins.[36]

All this packaging joined a growing stream of household waste that had to go somewhere. Food also joined the waste stream because refrigerators encourage people to buy more than they can use, which either spoils or ends up in discarded leftovers. Much food waste went into a device that increasingly was installed in new or renovated homes, the garbage disposer or "disposal," after General Electric's Disposall, which grinds up food and adds it to municipal sewage. Disposable items and packaging join discarded clothes, furniture, and appliances, and broken and unwanted things in trash bins. Modern consumers no longer make most everything they need and have lost the skills to make or repair things. The plastic of which things are increasingly made cannot be repaired in any case. Municipal or contracted garbage trucks collect all this and dump it in sanitary landfills. But there was not room for it all. By 1970, the flood of trash and garbage was filling landfills across the country and more and more communities built incinerators.[37]

Prosperity's environmental costs: waters

Dams proved crucial in the growth and spread of consumer capitalism. Dams control floods, provide irrigation water, supply hydropower, and facilitate economic growth and development. They also symbolize modernity, development, and state power used to benefit the people. The United States led a movement to build large dams in the 1930s with Hoover (or Boulder), Grand Coulee, and many other huge dams in the West and the Tennessee Valley. The Soviet Union followed the American example. A global dam mania took hold after the war.

Full of technological optimism that Western rivers could be engineered to promote economic and population growth, the United States embarked enthusiastically on two major dam-building sprees around the upper basin of the Colorado River, the only Western watershed with significant water. The Colorado River Storage Project, proposed in 1950, and the Central Arizona Project of 1968 led to construction of numerous dams, aqueducts, canals, and irrigation projects in five states.[38]

With its own confidence in experts' ability to create a new man and a new society and to control nature to boot, the Soviet Union embarked on postwar plans to replumb the nation's rivers, which tended to flow in inconvenient directions. Perhaps the greatest ecological disaster that Soviet engineering created was in the basin of the Aral Sea, a vast salt lake with no outflow fed by two long rivers with extensive, ecologically rich deltas. After Soviet water engineers were finished, endless fields of cotton grew on irrigated collective farms that had once been desert. Deprived of its water, the Aral Sea shrank. A salt desert replaced rich fisheries. Without annual floods, the river deltas suffered ecological collapse, desertification, aquifer depletion, and salt accumulation. Salt-laden dust storms now coat the countryside for up to three hundred miles and injure the health of crops, plants, animals, and humans. The climate for sixty miles around has become drier and warmer in summer and colder in winter. Heavy applications of agricultural chemicals under the Soviets also affected human and ecosystem health. The Soviets began construction of a water project to bring water from Siberian rivers to refill the Aral, but the enormous cost ended the scheme in 1986. With the end of the Soviet Union in 1991, the countries of Kazakhstan, Kyrgyzstan, Tajikistan, Turkmenistan, and Uzbekistan now divide most of the Aral basin. They broke up collective farms into subsistence farms and some large commercial farms, but they have only slowly begun to develop cooperative institutions to manage the water and revive the basin's ecological health.[39]

After World War II, the economic potential and powerful symbolism of dams appealed to ambitious leaders of rising and emerging nations in Asia, Africa, and South America. Gamal Abdel Nasser imagined greatness for Egypt with a dam at Aswan, for instance, which the Soviet Union was happy to build after the United States declined assistance in 1956. After its construction, electricity from the dam enabled substantial economic growth. Egypt benefited from electricity, reliable and steady river flows, improved navigation, and other things. Large landowners benefited most, growing cotton in the north and sugar in the south for export. However, the dam caused major environmental problems. As the river flow changed from seasonal highs and lows to a steady flow, farmers in the delta switched to corn (maize), whose low B-vitamin content spread pellagra. An ancient parasitic disease, bilharzia or schistosomiasis, broke out, spread by snails to bare-footed Egyptians walking in irrigation water. Treatment of irrigation ditches with copper sulfate and introduction of new drugs controlled the outbreak. In Lake Nasser, the dam reservoir, 20 to 30 percent of Egypt's yearly share of Nile water evaporated under the Sahara sun. Salinity rose and salt built up in fields irrigated with Nile water. Constant irrigation raised the water table so that monuments that have stood for thousands of years were damaged by water and salt. Starved of the rich silt of annual Nile floods, Egyptian fields require great quantities of fertilizer, production of which eats up much of the hydroelectric power. Lack of silt deprived nutrients to the eastern Mediterranean sardine fishery, which nearly vanished. Without annual loads of silt, the Nile delta is sinking and eroding. The silt builds up instead at the bottom of Lake Nasser. Siltation joins encroaching sand dunes from the desert to cut the reservoir's capacity.[40]

Global dam construction peaked in the 1970s, when 7,511 large dams were built. Most block rivers in India and China, which have together built over half of the world's 45,000 large dams. The hugest of all dams, holding back the largest of all reservoirs, is China's Three Gorges dam. Construction of the long-planned dam commenced in 1994 and was mostly complete a dozen years later. The dam's purposes were to allow ship traffic, control flooding of the Yangtze, and replace polluting coal powerplants with clean hydroelectric power. Three Gorges Dam's environmental costs (beyond the displacement of over eighteen million people and flooding of 1,300 archaeological sites) were immense. Inundated wetlands had offered habitat for rare or endangered species like the Siberian crane. Three Gorges played a role in the extinction of the Yangtze river dolphin. The reservoir also captured silt that formerly fertilized and built up the delta. Moreover, it sits atop a geologic fault. Dam failure would release catastrophe and death almost beyond reckoning. Finally, the huge reservoir does not have enough free space to retain large floods—floods in July 2010 overtopped the dam. On the other hand, its hydropower makes greenhouse-gas-producing coal power plants unnecessary, and the Chinese government has embarked on an extensive reforestation project in its watershed to slow silting of the reservoir.[41]

While dams and irrigation brought ecological derangements and inadvertent consequences to rivers and watersheds, rising postwar industrial production brought its eternal companion, rising water pollution. Multiplying population and expanding prosperity turned rivers around the world into foul, toxic sewers. After World War II, most North American, European, and Japanese towns and cities had waste treatment plants. Sewage treatment was too expensive for much of the rest of the world, where polluted water spreads disease and shortens lives.

Treatment of industrial waste is far more difficult. After 1945, it grew less possible for Europe and America to put off a solution. The war had exacerbated the problem of industrial pollution. Traditional industries had expanded rapidly to produce war material, focused on production above considerations like pollution. In the U.S., the southern shore of Lake Erie and the Ohio River downstream from Pittsburgh grew desperately polluted. Production and development of plastics and other new substances out of petroleum had also accelerated. Petrochemical plants concentrated on waterways near regions producing oil and gas. The area along the Mississippi River between New Orleans and Baton Rouge, Louisiana, known as the Chemical Corridor for its concentration of petrochemical plants, produced so much air and water pollution as to earn the alternate sobriquet "Cancer Alley."[42] Areas with both manufacturing and chemical plants hit rivers doubly hard. In the heavily industrialized Rhine River valley, for example, pollution was so bad that in 1971 a Dutch news photographer developed film in Rhine water. Many of the new chemicals and chemical waste products mimic hormones in humans. Ingested, they can cause cancer or affect sexual maturation.[43]

Items made from new plastics and artificial substances seemed to be everywhere in the years after 1945. Durable, cheap, and versatile, they appeared in toys, bags, cups, clothing, auto interiors, appliances, even plastic pink flamingoes to decorate yards. Instead of degrading, however, they break down into toxic chemicals and microscopic particles of plastic. When not disposed of properly, plastics get into waterways and the oceans, where they wreak havoc as organisms entangle in them, ingest them, and suffocate from them. As use of plastics spreads to poorer nations with inadequate waste disposal, they end up in the environment and the oceans in alarming quantities (currently an estimated ten million tons per year). Microplastics pass up the food chain in the flesh of marine creatures, including seafood. The sea salt and fish in your dinner put microplastic particles into your body.[44]

Prosperity's environmental costs: radiation

Around the world, countries both capitalist and Communist expected their energy demand to rise dramatically in the decades after 1945 and worried that supplies would fail. Dependence on unreliable foreign oil was a concern for many nations as well. Atomic energy, another product of World War II, promised a solution. Postwar prestige of scientists East and West was at its peak, so when scientific authorities promised "that our children will enjoy electric energy too cheap to meter," as Lewis L. Strauss, chairman of the U.S. Atomic Energy Commission, did in 1954, utilities and customers listened. The U.S. government created the AEC in 1947 to put atomic (or nuclear, after 1951) weapons and power under centralized, civilian control. But to increasing public distress, as the nuclear arms race with the Soviets sped up, atmospheric testing increased, and cancer-causing radiation blew around the globe. The AEC worried that sentiment might build to stop weapons testing, so it presented nuclear power as benign and beneficial. Still, nuclear power plants were so dangerous that the 1957 Price–Anderson Act shielded power companies from lawsuits over injuries caused by nuclear accidents. Because insurance companies would not insure plants, they could not be built without government protection and subsidies. France and Japan lacked energy supplies of their own and built plants enthusiastically. The Soviet Union aggressively built reactors. Many other nations also built power plants.

The AEC and its successor, the Nuclear Regulatory Commission, could not prevent negative stories and publicity. Evidence mounted that food was contaminated with radiation. Well-publicized accidents, like the Three Mile Island plant incident in 1979, and occasional catastrophes, notably at Chernobyl in 1986 and Fukushima in 2011, soured the public on nuclear power almost everywhere. Radiation released from bombs and accidents lasts for tens of thousands of years. Safe disposal of radioactive wastes is also extremely difficult. In addition, uranium mines and tailings expose workers and local people to radioactivity. Finally, nuclear fuel and spent fuel must be protected from theft, terrorist attack, or appropriation by governments for nuclear weapons.[45] Postwar government-inflated optimism about nuclear power collapsed in the 1970s and 1980s.

Consumer paradise lost

After the horrors and destruction of World War II, consumer capitalism promised to the world a new Eden, a garden of earthly consumer delights, where a veritable cornucopia poured out every good thing and many new and marvelous things. Hamburgers and french fries, cheap energy, miracle chemicals, plastic toys and flamingoes, cheap and abundant food, electric appliances and devices, computers, refrigerators, air conditioners, excitingly styled automobiles, democracy and prosperity for all—all poured out of consumer capitalism to a wondering world.

By 1970, a serpent had appeared in this garden. The biblical story in Genesis tells how the serpent was the first salesman, subtle and beguiling, promising great things, making forbidden fruit look good, pleasing, and desirable. In John Milton's *Paradise Lost*, Satan enters Eden in the form of a cormorant, symbol of greed. By 1970, consumers' eyes were opened, and they saw the curse that lay hidden behind those good, pleasing, and desirable goods. Population was growing breathtakingly quickly. Cities sprawled across former croplands. Water was unsafe to drink and air unsafe to breathe. Oil spilled and spread on the oceans. Wild species were declining. Poisonous chemicals covered farms and food. Soils were salinized and exhausted. Forests retreated. Deadly radioactivity permeated the world around us. In 1970, the curse mainly menaced America, Europe, and Japan, where the consumer capitalist Paradise had first established itself. Already, however, gardens of earthly consumer delights were beginning to spring up elsewhere, and the serpent slithered into every corner of the globe.

EIGHT

Selling Everything

The everything store

In the early twentieth century, industrial capitalism concentrated immense wealth in the coffers of corporations and the pockets of their owners. In 1901, J. P. Morgan created steelmaker United States Steel Corporation as the world's largest private corporate enterprise. A decade later, oil refiner Standard Oil[1] and at mid-century carmaker General Motors held that title.[2] Creation of U.S. Steel made Andrew Carnegie the world's wealthiest person, a place taken by Standard Oil's John D. Rockefeller a decade later. While GM made neither Henry Sloan nor anyone else the world's wealthiest, carmaker Henry Ford was the world's wealthiest businessman at the time of his death in 1947. Producing things could be very lucrative.

In the early twenty-first century, consumer capitalism enriches corporations and individuals of a different sort. In 2021, Amazon.com was the world's most valuable corporation[3] and its founder, Jeff Bezos, the world's wealthiest person. In contrast with U.S. Steel, Standard Oil, or GM, Amazon.com produces almost nothing. Heir more to the department store than the factory or refinery, the company serves as middleman between producers and consumers. Another middleman behemoth, Walmart, owned by the immensely wealthy Walton clan, earns more money than any other private company on earth. Other companies jostling Amazon .com at the top of lists of the most valuable public companies include Apple, Microsoft, Alphabet (Google), Tencent, and Facebook, which also chiefly offer services or intellectual property like software or video games, along with some consumer electronics. Uniquely among these companies, Apple earns most of its income from devices but, as markets for its products saturate and product sales make up a shrinking share of total income, Apple, too, looks to services as the growth sector of its future. In a sign of the mid-century energy transition from coal to oil, the notable exception to the trend away from producers is the continued presence of petroleum companies among top private corporations—among them Exxon, the direct descendant of Standard Oil of New Jersey, one of the corporations created when an antitrust decision broke up Rockefeller's Standard Oil in 1911.

In contrast to Carnegie, Rockefeller, or Ford, industrial production did not make rich men of Bezos or his fellow super-billionaires Bill Gates of Microsoft, Mark Zuckerberg of Facebook, Larry Page and Sergey Brin of Alphabet, the Walton family of Walmart, or Warren Buffett of investment firm Berkshire Hathaway. (Producers still exist, like Elon Musk of Tesla and Bernard Arnault of the French luxury goods company LVMH, but they are a shrinking minority.) Bezos had risen from Wall Street in the manner of J. P. Morgan or William Durant. But while Morgan and Durant used their Wall Street experience to create the holding companies U.S. Steel and GM to buy and coordinate manufacturing companies, Bezos left Wall Street to launch a company that owned little more than computers and warehouses and sold items through a revolutionary new medium, the Internet—which, ironically, in the 1990s predictions claimed would eliminate middlemen.

As 1920s-style consumer capitalism aged, the nature of business corporations changed and further accelerated consumption and environmental problems. The form of consumer capitalism that had made corporate giants of GM and Ford ceased to exist. A new era for capitalism and for the environment began around 1970 and brought forth Amazon.com and the other colossal global corporations. Rather than manufacturing items for retailers, the leading enterprises of the twenty-first century sell to consumers and sell enormous amounts very quickly. The salesman, heroic servant of the Roaring Twenties economy, has taken command. Economic growth under consumer capitalism depends upon moving cash through the economy with ever-increasing velocity.

Some scholars have called the postwar age a "Great Acceleration."[4] After 1970, the Great Acceleration accelerated even faster. The environment, however, can no longer bear the cost. The relentless, rising torrent of consumer goods that gives Amazon.com its apt name places unabating demand on extractive industries for resources and energy. Another "Amazon River" of waste flows into air, water, and land. Forests, species, and soil dwindle, everlasting toxics and plastics accumulate, and the climate changes with a rapidity that awes researchers. From the deepest ocean trench to the upper atmosphere, consumer capitalism has deranged the entire earth.

Bezos and the new consumer capitalism

Bezos conceived of Amazon.com as a scheme to get rich. He was a vice-president at the Wall Street firm D. E. Shaw. Founder David Shaw had left a professorship in computer science at Columbia University in 1986 to work for Morgan Stanley, the investment bank founded by J. P. Morgan's grandson. Two years later he founded D. E. Shaw, an early quantitative hedge fund, which invests according to mathematical or statistical formulas rather than human judgment. Shaw devised proprietary computer algorithms on high-speed computers to find and profit from patterns in global financial markets. He hired the brightest and most innovative computer science graduates of the best schools. Bezos graduated from Princeton University in 1986 with a degree in computer science and moved restlessly from one Wall Street firm to another until Shaw hired him in 1990. Bezos thrived and rose quickly through the ranks.

Shaw kept an eye on the early development of the Internet. He foresaw its commercial potential earlier than most investors. Shaw regularly discussed possible Internet ventures with Bezos, which Bezos would research. Bezos noticed that the amount of information sent over the Internet had multiplied an astonishing 2,300 times in a single year, 1993. This nearly miraculous growth suggested almost unfathomable potential. One idea the two men discussed was an "everything store" that would sell every kind of item worldwide. Since no company could immediately start selling everything, Bezos decided that the way to start the store would be to sell one product and grow from there. He made a list of potential products that could profitably be sold online, of which books looked most promising.[5]

Bezos left D. E. Shaw in 1994 and moved to Seattle, in large part to avoid sales tax except for sales in the lightly populated state of Washington. He started the online bookseller that would become Amazon.com, named for the river with the largest volume, an ambitious but prescient choice. Bezos worked unrelentingly to make Amazon.com "get big fast" so that it could stay ahead of competition. (Tellingly, "Relentless.com," was one of the names that Bezos considered for his new company before he settled on Amazon. Typing "Relentless.com" in a Web browser's address line still redirects to Amazon.com.) Bezos hired the smartest, most creative people he could find. He drove them to innovate quickly to push ahead of Microsoft, Barnes and Noble, Walmart, and other large established companies, which, invested in older ways of doing business, were slow to understand commercial possibilities of the Internet. Amazon.com's employees accomplished much as it grew from an operation selling out of a garage to global vendor. They learned how to process online payments securely. They developed the "1-Click" technique that applies address and payment information from a previous purchase to a new one without customers having to retype it. They developed a program that used a customer's sales and browsing history to recommend books to buy. When Bezos realized that Americans did not read a lot of books, which limited expansion opportunities, Amazon.com began to add its immense array of

Figure 13. Álvaro Ibáñez, Una visita al gigantesco Centro Logístico de Amazon España en San Fernando de Henares (Madrid), 2013. An Amazon Fulfillment Center in Madrid is stocked with thousands of consumer items waiting for workers to quickly and efficiently send them to our doorsteps. The aptly named "fulfillment center" fulfills consumers' orders. Yet we also often find personal fulfillment through consumer goods. (Álvaro Ibáñez via Wikimedia Commons: https://commons.wikimedia.org/wiki/File:Amazon_Espa%C3%B1a_por_dentro_(San_Fernando_de_Henares).JPG)

other products. In 2006, Amazon.com began to sell access to its storage, computing, and database computer systems, so-called "cloud" services, which has become a very lucrative business. One day, perhaps, Amazon .com will indeed sell everything.

Amazon.com completely upended established businesses and retail establishments. Bezos was a pitiless competitor. Algorithms and programs replaced salesmen. The company challenged and changed publishing, since, ironically, antitrust law prevented publishers from forming a united front against it. As its offerings expanded beyond books, Amazon.com drove local retailers out of business all over the country, from small bookstores to specialty shops, from bookstore chain Borders to that venerable pioneer of mail-order sales, Sears, Roebuck.

If Amazon.com and companies like Walmart, Apple, Google, and Facebook represented the shift in consumer capitalism from the days of U.S. Steel and General Motors, their founders represented a new kind of entrepreneur. Like the McDonald brothers and Ray Kroc, few grew up in the New England or Reformed Protestant traditions that shaped industrial capitalism. Rather, they were raised in the cultures of groups formerly associated with the old Democratic Party, that is, Protestant Southerners and Catholic and Jewish immigrants, whose cultures tolerated greater ease with the self and its indulgence, less guilt over wealth and self-gratification, and less moralism. (Exceptions include former Congregationalist Bill Gates of Microsoft and former Presbyterian Warren Buffett, both of whom have dedicated themselves to giving their fortunes away, much in the mold of Carnegie or Rockefeller.)[6]

Bezos's mother was raised a Texas Methodist and his adoptive father was a Jesuit-educated Cuban Catholic immigrant and petroleum engineer for Exxon. Born in 1964, the very bright, very driven, and very competitive young Bezos enjoyed an affluent suburban childhood in Texas and Florida. Although he has given scant evidence of any religious interest, Bezos may identify as Catholic. From his stepfather, Bezos acquired a strong dislike for government interference with business and personal affairs. As one of the company's initial investors observed, "Jeff Bezos is a straight-up libertarian . . . [with] the canonical neoliberal perspective: that the only purpose of corporations, the only purpose of shareholders, is to enrich themselves to the exclusion of everything else . . . Maximize shareholder value and somehow magically that will create the common good."[7] Libertarian or neoliberal values harmonize beautifully with values that consumer capitalism celebrates: individual pleasure, self-gratification, entertainment, and consumption. Bezos was still very young when he formulated two life goals: to get rich and to go into outer space. Amazon.com had achieved for him the former. In 2000, he founded a private spacecraft company, Blue Origin, that realized the second for him in 2021.[8]

The death of the Future and fulfillment, identity, and meaning in consumption

The transition from the 1920s-style consumer capitalism to the consumer capitalism of Amazon.com and Walmart took place in the years around 1970. Culture, society, politics, and economy were changing in ways that could be felt. The winds shifted, the skies darkened, and the world grew fearful of rather than hopeful for the future. In the postwar era, ideas and movements for a better world had abounded. Around 1970, idealistic, optimistic social and political movements veered into frustration and violence. The American Civil Rights movement, demonstrations against the Vietnam War, and campus protests were taken over by violent extremists or met with violent government suppression. Right-wing military coups with American support overthrew democratic governments in Brazil, Greece, Chile, and Argentina. 1968 was the year of violently suppressed risings and revolutions from Paris to Prague. German and Italian far-left groups began long campaigns of violent activities. Idealistic Chinese youth in the Red Guard spiraled viciously out of control as they chased phantom enemies of the revolution. In Europe and Latin America, campus protests also descended into violence. In Africa and Asia, often hasty and disorganized decolonization fostered political confusion, unrest, rebellion, and civil war.

New forces weakened the cultural brakes on the advance of consumer capitalism. Commercial television inculcated consumer values in the first generation that grew up in front of a television set. The popular human-potential movement proposed that the self required no control or suppression for the common good but rather fulfillment to realize or actualize its potential.[9] Introduction of the birth-control pill in 1961 freed young people from the fear of pregnancy, unleashed a "sexual revolution," and propelled steady sexualization of culture and media.

The religious basis of western European and North American culture shifted as consumer capitalism and social forces eroded communal traditions and encouraged individual eclectic spiritual experimentation. On both sides of the Atlantic, young people abandoned traditional parish-based denominations. Religion became an item for consumers to pick and choose according to their personal desires. Americans now "shop" for religion and churches. Buddhist, Hindu, and indigenous spirituality (or their popularized counterfeits) attracted many. Denominations that emphasized the individual's relationship to and experience of the divine grew rapidly, especially Pentecostalism, an ecstatic branch of Protestantism born in Los Angeles in 1906, which deeply influenced other churches and popular culture.[10] After 1970, Pentecostalism also spread rapidly in Latin America and Africa. All churches now offered a therapeutic spirituality that promised to make believers thinner, better, richer, and happier.[11]

Consumerist cultural currents also carried liberal and leftist politics into new directions. The American Democratic "Roosevelt coalition" of non-Protestant immigrants, Southern whites, and African-Americans turned progressivism and liberalism away from a relatively coherent Progressive social vision of northern Reformed Protestantism. After 1970, Reformed Protestants would disappear from liberal leadership altogether. Liberalism moved toward greater emphasis on individual rights and identity. White Southern evangelicals left the Democratic Party and after 1980 formed a reliable voting block in the Republican Party. Historically, white Southern evangelical Protestantism had promoted little social activism and supported slavery, segregation, and white supremacy.[12]

After 1970, cultural forces weakened the West's confidence in itself. Europe lost the will and power to maintain its global empires, while America's rise diminished its sense of self-importance. Appreciation for and interest in foreign and native religions and cultures grew in parallel with a growing discomfort with Western religions, institutions, and even science. Parties of the liberal, labor, and social democratic left faltered in promoting their principles and lost political ground. Cynicism and libertarian or neoliberal ideals eroded reforms of the previous century, especially in the U.S. and Britain. Marxist self-styled "scientific socialism" weakened and its influence waned. Marxism's oldest and strongest state, the Soviet Union, and its allied nations grew slowly weaker and collapsed suddenly in 1991. Marxist movements faded into irrelevance but held on in China, North Korea, and Cuba, although in China with pragmatic, cautiously market-oriented policies.

The emotional, irrational forces of conservative religion surged into the vacuum left in the wake of retreating democratic, socialist, and religious ideals born of Enlightenment confidence in human reason. Christian, Jewish, and Muslim fundamentalisms swelled in popularity and political influence in the U.S., Israel, Iran, Saudi Arabia, and elsewhere. Oil money funded religious extremism in both America and the Islamic world. As faith in the future faded, retrospection and nostalgia pervaded the post-1970 era. Many people worldwide sought to hold on to a mythical past. From the Industrial Revolution up through the 1960s, people often eagerly speculated about the future and what wonders it would bring. The year 2000 arrived without any grand forecasts of the great and glorious century or millennium that awaited. Cynicism and pessimism dominate the mood.

People everywhere now find meaning and express their identities through shopping, buying, owning, and consuming.[13] With ubiquitous advertising, significant cultural undermining of intellectual and moral foundations of progressivism and leftist politics, and expanding political and cultural propaganda networks funded by corporations and wealthy individuals, consumer capitalism insinuates itself everywhere. Those who came of age after the 1970s, like Bezos, Zuckerberg, and other Internet moguls, absorbed this libertarian ethos.

Slowing economic growth and technical innovation

Around 1970, a century of astonishing economic growth in the United States and technical innovation around the world came to a close. In Europe and Japan, growth continued for a while because their economies had trailed the U.S. until the 1970s. Starting from a much lower economic level, China's economy began to grow at astonishing rates in the 1990s, tapering off in the 2010s.

Technical innovation since 1970 has created no revolution in daily life like that of the century after 1870 and especially the half century after 1920. Recent innovation has been confined almost exclusively to communication, entertainment, and information technology. Almost all other fields merely advanced on earlier innovations made between 1870 and 1970. It would be far easier for us living in the West today to go back in time to live in 1970 than for either us or someone living in 1970 to return to 1870, or even 1920. Life a half-century ago would be different, of course, but rather familiar. We would miss smartphones, personal computers, the Internet, and microwave ovens. Nevertheless, already in 1970, men walked on the moon, satellites of all kinds orbited the earth, and familiar appliances filled homes. Almost all homes boasted electricity, gas lines, central heating, municipal water, water heaters, bathtubs, and flush toilets. Plastics and artificial fibers appeared commonly in consumer goods in shopping malls. Televisions, movies, and recorded music entertained. Antibiotics and vaccines controlled infectious diseases and hospitals treated patients with familiar drugs, chemotherapy, and radiation. Cars and trucks drove modern highways lined by franchise restaurants selling fast food in disposable containers and cups. Jetliners crossed skies around the world. Diesel locomotives pulled trains, supertankers carried oil, and container ships plied the seas. People used credit cards to buy processed food at supermarkets and ready-made clothing at malls and department stores. A tiny handful of those things was available to some people in 1920. Few had any of them in 1870.[14]

American manufacturers' postwar dominance had made them complacent. They failed to keep the rate of investment in capital stock high while European and Japanese industries rebuilt and began to benefit from American-style consumer capitalism. The experience of U.S. Steel was typical. With infrastructure complete, domestic steel demand slowed but rising foreign competition tightened foreign markets. The 46.6 percent American share of the global steel market in 1950 slipped to 26 percent in 1960 and twenty percent in 1970. In 1970, Yawata and Fuji merged into Nippon Steel, and took the title of world's largest steelmaker from U.S. Steel.[15]

The American economy stagnated in the 1970s. In 1971, the U.S. ran its first trade deficit in a century. Manufacturing's share of the economy shrank as the service sector grew. Between 1973 and 1981, the growth of the labor force in eating and drinking establishments exceeded the number of employees in the auto and steel industries together. Pay and benefits for service employees failed to match those of factory union workers.

The Dow Jones Industrial average broke 1,000 for the first time in 1972 and then in 1973 and 1974 lost almost half its value, with parallel crashes on European exchanges. Middle Eastern crises in 1973 and 1979 caused two oil shortages that hurt the economy and worsened inflation. Detroit automakers lost market share as rising gasoline prices turned consumers to smaller, fuel-efficient foreign vehicles.[16]

Governmental economic policies and consumer capitalism

Changes in government economic policies after 1970 also guided the transformation of consumer capitalism. Libertarian-tinged economic theories, like Bezos's almost mystical belief that free markets would solve all problems, influenced the American government in dealing with both unemployment and high inflation. Hoping to boost manufacturing and foreign demand for American goods, in 1971 Nixon ended the Bretton Woods agreement that had stabilized international financial markets since World War II. Currencies floated against each other, prone to speculation and fluctuations in no country's best interests. Cash and capital flowed across borders following highest short-term profit rather than long-term investment needs, creating an international capital market and making possible such hedge funds as D. E. Shaw. To fix governmental economic regulations that either did not work as planned or were outdated, administrations of the 1970s and 1980s proposed, not better regulation, but no regulation. The Democratic Congress deregulated natural gas prices, the trucking industry, railroads, airlines, and banking and weakened regulation of other industries. Deregulation left the economy less stable. Financial crises have regularly hit the country as they never did between 1945 and 1970. After 1980 America and Britain led the way in dismantling the social safety net and weakening unions, which had boosted prosperity and promoted income equality. Blue-collar wages and middle-class income ceased to rise while tax cuts and deregulation sent the wealth of the richest soaring, setting the stage for the rise of today's Internet billionaires.

Cash from tax cuts which ought to have gone into investment in capital stock or manufacturing went instead into financial markets. U.S. Steel, for example, did not use its cash to modernize aging plants for the global marketplace. It closed them and invested in diversifying its holdings. In 1986, the company dropped "steel" from its name to become USX.[17] This process of financialization, in which corporations invest capital in profitable companies instead of in manufacturing, grew commonplace. Giant corporations and hedge funds bought up smaller, profitable companies. Industries from food to pharmaceuticals to publishing concentrated under control of ever fewer corporations, which accumulated profits and economic and political power. Antitrust regulators looked the other way if consumers were not directly hurt.

By the 1980s, many companies found themselves vulnerable to a financial sector looking for profits, not improved production or a better society. Corporate managers (as well as Internet billionaires like Bezos) sought as much as possible to raise stock prices and "shareholder value," a phrase coined in 1983. Concern for shareholder value had never before guided management, which had usually focused on corporate value to society or on long-term growth strategies. Now management's primary fiduciary duty was to shareholders. Money flowed to banks and investor groups. Corporate raiders, private equity funds, and hedge funds raised money through high-interest bonds ("junk bonds") to buy companies, sell off their assets to repay bondholders, and then, hopefully, profitably sell the more-valuable stock of the shrunken, dismantled company.

General Electric's chairman Jack Welch was a pioneer, widely admired for financializing GE to maximize shareholder value. Raised by poor Irish Catholics in the Boston area, with no particular talent for engineering or design, he rose quickly through management to lead the company from 1981 to 2001. The company grew even as it aggressively downsized, a word coined in 1982 to describe firing employees to improve financial performance. GE lost 100,000 employees. Research and development was gutted. Companies that followed Welch's influential lead included Boeing, whose engineering-oriented culture changed to a financially oriented one. It did not last. After Welch retired, stock price and company value plunged, and GE was removed from the Dow Jones Industrial Average in 2018. Meanwhile, Boeing produced the flawed 737 MAX, which twice, in 2018 and 2019, crashed loaded with passengers. Sales froze and investigations uncovered serious systemic issues in Boeing's organization.[18]

These changes in governmental policy, corporate governance, and financial markets has made growth under consumer capitalism more fragile. Republicans and sometimes Democrats have passed deregulation and tax cuts for the wealthy and corporations. This economy disproportionately fills the pockets of brokers and speculators. Companies increasingly compensate executives with stock options, which leaves them with even greater incentive to act with an eye to stock prices. In the 1990s, the stock market soared to unheard-of heights before Internet and technology stocks led to a crash in 2000. Predictably, deregulation led to risky banking and financial behavior, which created an American housing bubble. The bubble burst in 2008 and very nearly brought the global economy down.[19]

Perhaps one of the forces keeping economic growth below pre-1970 levels has been lack of cash. Crucial to robust consumer capitalism are lower and middle classes, who spend a far greater share of income on consumer goods than the wealthy. Their income has stagnated and the income of the poorest has declined. The poverty rate, which had been at the historically low level of eleven percent in 1973, has risen steadily since. Other elements that stoked economic growth have also suffered under the rising political power of libertarian and neoliberal billionaires and corporate leaders. Under pressure to privatize education from free-market-oriented private foundations, especially the Bill and Melinda Gates Foundation and the Walton (of Walmart) Family Foundation, investment in public education has suffered.[20] Keeping taxes low, government neglected infrastructure. Younger people in wealthy countries around the world marry later and have fewer children, which means populations are aging, a problem that large immigration into the U.S. has disguised since the 1980s. Fewer workers will have to support a growing aging population. Young Americans also must contend with surging educational and consumer debt, which along with stagnant income has made them the first American generation poorer than its parents. Further drags on the American economy include the large number of single women with children who cannot escape poverty, along with the absurdly large number of incarcerated young men, especially young black men.[21]

In the 1970s, government regulations, thinning profit margins, and economic frustration prompted conservative business leaders and wealthy investors to organize politically, not as a party, but in an interlocking system of lobbyists (who by 1980 outnumbered federal employees), think-tanks, business groups, and "free-market" endowed chairs and institutions in universities. When in 1976 the Supreme Court ruled that political donations were constitutionally protected speech, campaign contributions rose significantly. In a second decision in 2010, the Court took the lid off corporate campaign contributions. Since then, corporations have poured vast amounts into anonymous "dark money" groups, political action groups, and political campaigns. Not only has this pushed politics in a strongly anti-government, anti-regulatory direction, but has influenced the Supreme Court to the degree that it usually smiles upon corporate plaintiffs and blocks government regulation.

Since parallel patterns have emerged in other developed nations, especially the Anglophone countries, all these changes have damaged global consumer capitalism. As capital globalized after 1970, so did manufacturing and trade, sped by cheaper transportation. Every industrializing nation has always aped the leading industrial power of its day, to eventually surpass it due to cheaper labor, improved and more efficient manufacturing, or both. After World War II, American manufacturing found itself challenged first by Europe, then both Europe and America by Japan, South Korea, Taiwan, and finally China. Sweatshop industries sprang up in poorer nations like Bangladesh and Vietnam. With its huge, cheap labor force, China flooded world markets with inexpensive goods. American and European investment money flowed abroad. Corporations outsourced operations overseas. American factories closed. Workers lost well-paid union jobs and joined the low-paid service sector. Only American and European industries that added high value or required skilled labor survived.

Agriculture and the food industry

Consumer capitalism transformed the food industry after 1970. Factory-style agriculture fed food corporations, which filled supermarket shelves with packaged processed products. A few mostly American giant companies took over much of the world's food chain. "Big Food" created a very profitable system in the United States that it exported to countries around the world. It also exported the "American diet" of bottled sodas and processed foods. American health problems followed, especially obesity, which is a problem nearly everywhere now.

Government policy pushed American farmers toward consolidation. In the United States, government policy under the Republican administrations of Dwight Eisenhower, Nixon, and Reagan shifted government support from family farms to agribusiness. Farms consolidated, until fewer than ten percent of farms earn more than 85 percent of farm income. Traditional diversified family farms could not survive. The smallest 80 percent of farms now produce too little income to support families, so that farmers or spouses need second jobs. Demand for farm labor continued to decline, except for seasonal migrant laborers harvesting fruits and vegetables. Between 1930 and 2000, farm population dropped by a third while production on the same acreage doubled. Farm income also rose. While rural poverty among sharecroppers and tenants was a real problem in 1945, it had almost vanished by the twenty-first century, although poverty remains a terrible problem for migrant workers. European Union policies have promoted parallel changes across Europe.

American farms produce tremendous amounts of food very cheaply. For farms to produce so much at such low cost, farmers need to apply artificial fertilizer, pesticides, herbicides, and fungicides on every available acre. In the laboratories of chemical conglomerates, researchers modify genetic codes to create strains of plants that can resist pests and chemicals. Farmers specialize in a single crop. Large, endless fields of corn, cotton, wheat, or other monoculture cover many regions of the United States. In the Southern states, producers developed concentrated animal operations, which crowd thousands of chickens, pigs, or cattle together in rather cruel arrangements, to be fed until slaughtered or, for chickens, kept for their eggs.[22]

The history of postwar agriculture stands Malthus on his head. As Euro-American farming techniques spread around the postwar world, so did its productivity. Global population increased 2.4 times between 1950 and 2000, but agricultural production tripled. Unneeded agricultural laborers have crowded into cities in China, India, Africa, and Latin America to provide labor pools for factories, mines, and other nonfarm sectors of economies. World trade in agricultural commodities like bananas, coffee, cotton, beef, and palm oil rose. Before World War II, agricultural output accounted for about fifteen percent of global trade but, by the end of the century, a quarter to a third of it. Varieties of fresh fruit and vegetables at European, American, and Japanese supermarkets multiplied. European and American immigrants hungry for familiar agricultural products have pushed further diversification of the produce section.[23]

The computer revolution

Without any doubt, the most revolutionary economic development of the post-1970 era, without which Amazon.com could never have existed, was development of the computer technology industry. This manufacturing sector grew out of innovations by or funding from the U.S. military during World War II and the Cold War. Stanford University in Palo Alto, California, was poised to benefit from its superb engineering school and proximity to military installations, to the NASA Ames Research Center, and to a nascent electronics industry. Nearby electronics companies included Xerox Corporation (with its research center in the university-owned Stanford Research Park), Hewlett-Packard, Fairchild Semiconductor (which invented silicon semiconductors, sold at first mainly to the military), Intel, and video-game-maker Atari. Early venture-capital firms funded by former electronics employees sprang up nearby as well. Between 1968 and 1969, the U.S. Department of Defense's Defense Advanced Research Projects Agency developed ARPANET connecting computers at Stanford and other universities, which Stanford engineers took the lead in developing into the internetwork, or Internet.

The concentration of silicon-based semiconductor technology earned the region the sobriquet "Silicon Valley" in 1974. Silicon Valley derived driving energy from a liberal-leaning libertarian culture of the region. Nearby San Francisco was the birthplace of the Beatnik movement of the 1950s and the hippie counterculture of the 1960s (fueled by psychedelic drugs tested on Stanford student volunteers), and a major center for the burgeoning gay rights of the 1970s. Amateurs and engineers alike hoped to break the corporate hold on computer technology and democratically give its power to everyday people. Later, they hoped the World Wide Web would usher in an era of individuals connecting with each other and expanding freedom, communication, and sharing of information without need for corporate middlemen. Ironically, of course, many went on to found giant corporate middlemen profiting handsomely from collecting personal information about computer users and their usage of computers and the Internet.[24]

Anticorporate idealism produced the personal computer. Like the steam engine, light bulb, automobile, and airplane, it emerged from the work of individuals tinkering in sheds or garages. Xerox had developed the first personal computer but never commercialized it. In Palo Alto, a community of young engineers was experimenting with computer technology. In 1976, two of them, Steve Jobs and Steve Wozniak, founded Apple Computer and borrowing or stealing ideas from Xerox and Atari brought out the Apple II in 1977, the first personal computer. In 1979 in Seattle, Washington, Bill Gates bought a computer operating system, which he renamed MS-DOS, and sold it to IBM to use in the personal computer it introduced in 1981. The World Wide Web appeared in 1991, the first browser in 1993, and, at the same time, the first search engines. More and more people bought computers, sent e-mails, and "surfed" the Internet. D. E. Shaw and Bezos were among the early figures looking for ways to profit from the new Internet. A rush soon followed them that pumped up an investment bubble. Amazon.com was one of the companies to survive when the bubble burst in 2000.[25]

The shine on the computer age tarnished quickly. Businesses expected computers to yield greater productivity. But aside from the decade after 1994, productivity has continued to grow more slowly than the postwar era before 1970. Beginning in the 2010s, online retailing began to make dramatic inroads into the business of what were now called brick-and-mortar stores. In the 1980s and 1990s, Walmart, K-Mart, Target, Home Depot, Barnes and Noble, and a host of other "big-box" stores had revolutionized shopping and retailing. Traditional department stores, even venerable J. C. Penney and Sears, struggled against them. Big box stores offered cheap prices but paid their insecure, part-time workforce such low pay with so few benefits that they often qualified for public assistance. Many small independent stores went out of business. Amazon .com and online shopping put big box stores on the defensive, with low overhead, unmatchable selection, low prices, ease of purchase, and ever-faster delivery. Streaming movies and television series from Netflix and other companies eroded audiences for movies and television programs. Rapid development of smartphones in the new century made personal computers remarkably portable and brought the entire Internet into people's pockets and purses. Reaching consumers with advertising, data-collection, and one-click purchasing was never easier.[26]

Figure 14. Mobile payment terminal, in Fornebu, Norway, operated by near-field communication technology from Norwegian telecommunications giant Telenor, 2011. The ubiquitous smartphone can now be used to make purchases without the need for cash or credit card, making consumption astonishingly easy. It also perhaps encourages purchases by distancing the act of buying from its monetary cost, unlike payment by cash or even credit card. Consumerism has never been so seductive. (HLundgaard via Wikimedia Commons: https://commons.wikimedia.org/wiki/File:Mobile_payment_03.JPG)

Commercialization of the Internet sped up and expanded the reach of consumer capitalism. However, in it the tech industry had created a monster. To offer everything free, companies like Alphabet with its Google search engine and "social media" like Facebook collect immense amounts of often very personal information to sell. Advertisers then use the information to tailor and target advertisements to just the people likely to find their products interesting. The information is also useful to target people for political advertisements and messages. Unfortunately, the wide-open nature of the Internet, once touted as liberating, encourages criminal activity, malicious actors and software, pornography, political propaganda, and false, misleading, and malicious information.

The Great Acceleration accelerates the global environmental crisis

The rising tide of environmental problems in 1970 became a tsunami of troubles a half-century later. Rapid growth of population and consumer capitalism pushed demand for extractive commodities ever higher and dumped ever more waste products into air, water, and soil. Industrial capitalism abandoned its birthplaces in Europe and the U.S. China and India, two rising countries with vast pools of cheap labor, took advantage of falling costs of importing raw materials and exporting parts and finished goods to adopt their own versions of industrial capitalism. Old environmental woes followed the movement of industry to new lands.[27]

To feed consumers and provide them with food products they desire, manufacturers need agricultural commodities. Food giants Kraft Heinz, General Mills, Conagra, Unilever, and Delmonte effectively control the global food commodity chain. They tell farmers what to plant and how much they earn. Through ubiquitous advertising and promotion they tell consumers what to eat and how much to pay. Multinationals Tyson, JBS (a Brazilian company), Cargill, and Smithfield (owned by the Chinese WH Group) play a similar commanding role in the beef, pork, and poultry industries.[28] Midwestern American farmers have found themselves locked into using substantial inputs of artificial fertilizers and chemical herbicides and pesticides to raise genetically modified corn. Some corn feeds and fattens livestock. Most goes to processors, who turn the corn into a huge array of products: starch, oil, high-fructose syrup, emulsifiers, stabilizers, and viscosity-control agents for thousands of processed food products, plus ethanol for cars.[29]

In the tropics, agriculture drives rainforests back. Deforested regions of Brazil's Amazon have become soybean fields for cattle feed or cattle pastures to make hamburgers for Europe and North America.[30] Forests in Indonesia, Malaysia, Nigeria, Thailand, and Colombia increasingly make way for palm-oil plantations. Production of palm oil quadrupled between 1995 and 2015 and is expected to quadruple again by 2050. This versatile oil yields about 200 products: cooking oil, ingredients in baked goods and foods low in saturated or trans fats, detergents, liquid soaps, shampoos, cosmetics, food preservatives, adhesives, personal care products, and biofuel.[31] Tropical banana plantations have replaced forest and produce more than any other fruit in the world, driven by urbanization and urbanites' desire for a reliably tasty fruit. Prone to pests and diseases, bananas require large applications of chemical pesticides, herbicides, and fungicides, which leach into water tables, damage offshore coral reefs, and harm worker health. Soil left bare by herbicides erodes in tropical rains sending chemical-laden silt to clog streams. Soil depletion or presence of fungus regularly forces growers to abandon plantations and cut new ones out of forests.[32]

A large proportion of the protein needs of the world's growing population comes from the sea. The biomass of its larger animals has dropped radically. Old, heavily exploited fishing grounds like the North Sea and Baltic and the northwest Atlantic had suffered for centuries from declining fish stocks. The problem worsened as technology improved in the nineteenth century with the introduction of steam power. Still, scientists' and fishermen's postwar optimism for the sea's potential was high—*The Inexhaustible Sea*, as the title of a 1955 book had it.[33]

Whaling first raised international alarm at disappearing ocean species. In the late nineteenth century, Norwegians developed efficient whaling factory ships with deadly explosive harpoons fired from cannons, with flotation technology to keep carcasses from sinking. Although kerosene lanterns gave whales a reprieve when they replaced whale-oil lamps, whale oil has unique properties as an industrial lubricant. Before the late 1960s, the transmission fluid of every car with automatic transmission had sperm-whale oil in it. Whale oil lubricated intercontinental ballistic missiles.[34] Public outcry in the 1970s over the global decline of whale species finally pressured most nations to cease or drastically cut back whaling. After much research, automakers synthesized a replacement for whale oil. Since the 1970s, the number of whales has increased steadily but remains far below their population in 1850.[35]

Much more significant have been the technological changes that threaten to sweep the seas clean of practically all fish. By the 1960s, the British had developed huge trawlers with freezer compartments that allowed fishermen to range far into the Arctic, away from ancient, depleted fishing grounds of the North Sea. The trawlers had stern ramps to pull nets aboard and equipment to process cod liver oil and fish meal. Eastern European nations developed floating fishing villages. With onboard medical staff and crew entertainment to lessen the need to return to port, giant factory ships today cruise distant waters with fleets of smaller trawlers to bring fish to mother ships for processing. Similar distant-water fleets from Japan and Taiwan have joined them. With sonar to find schools of fish, these fleets suck the life out of the sea. Development of new bottom-trawling technology and nets made of new kinds of materials has allowed fishermen to capture entire schools or leave the seafloor a lifeless desert. Ships toss dead or dying turtles, sea mammals, sponges, corals, and anything else uncommercial back into the sea as "bycatch." Thousands of miles of abandoned nets and fishing gear drift and kill long after ships have gone home. Coastal pollution has also killed off fish, as have the growing dead zones from eutrophication caused by nitrogen in sewage and fertilizer runoff.[36]

Biodiversity on land faces dire threats from several different directions. Expansion of commercial plantations, farms, and pastures has destroyed forests and taken habitat for many forms of life. Collectors, hobbyists, and exotic game ranches drive demand for many species of birds, reptiles, mammals, and fish, especially rare ones. Asian walking catfish in Florida, Japanese snow monkeys in south Texas, and many other non-native species have escaped or were released in new lands. Occasionally, like the Burmese pythons in the Florida Everglades, they have wreaked havoc in local ecosystems. Some deliberately introduced species, often with the best of intentions or scientific support, become invasive and transform and occasionally upend ecosystems. Examples include the North American prickly pear cactus in India, the American gray squirrel in Britain, or Asian carp in the United States. Finally, as Chinese become more prosperous, demand for luxury goods of ivory and for traditional medicines have driven an illegal international trade that has pushed elephants, rhinoceroses, tigers, and other animals to the edge of extinction, or over it.

Cheap global transportation has introduced some of the most troublesome invasive species. In the cargo and ballast water of ships, in the holds and cabins of aircraft, on travelers' shoes and clothing and in their luggage and bodies, and in crates and shipping containers, hitchhikers and stowaways from microbes to seeds to sea life to insects and small animals move from continent to continent every day of the year. Some blend innocuously with local biota. A terrible few work vast destruction, such as the fungus that destroyed the great chestnut forests of North America in the nineteenth century or the fungus that even now is wiping out the world's amphibians. Between species extinctions and species transfers both purposeful and inadvertent, humans have remade and are still remaking global ecosystems.[37]

Throwaway capitalism

To walk into a big box store like Walmart or Ikea is to step into a cornucopia of inexpensive wares. Planned obsolescence has reached such things as clothing and furniture, which former generations wanted to last. They are cheap and designed to be soon discarded. Fast fashion's combination of high speed, low cost, low price, high profits, and high environmental impact perfectly illustrates the drive of consumer capitalism to profit by speeding the passage of money from one pocket to another at the expense of the natural world. Fast fashion sends clothing through the supply line and retail floor to customers at a lightning pace. Americans buy five times more clothing than they did in 1980 and wear the average piece of clothing seven times.[38] The consumer fashion business creates environmental problems on its entire journey from seed (or chemical plant) to dump. It uses perhaps one-quarter of the world's chemical production in raising and processing cotton or making artificial fibers and in treating and dying cloth. When Americans have tired of a piece of clothing, if it does not end up on a charity shelf or in Africa (Kenya alone receives 110,000 tons annually), they dump each year tens of millions of tons of clothing in landfills, about twice as much as twenty years ago. Manufacturers and retailers bury, shred, or incinerate the twenty percent of clothing that never sells.[39]

Add, too, the proliferation since 1970 of disposable products. Many disposables had been invented in the 1920s or the nineteenth century, but ingrained thriftiness and resistance to wastefulness limited their distribution, an attitude that the Great Depression and wartime rationing reinforced. Postwar prosperity eroded that resistance. Advertising promoted the values of "freedom" and "convenience" instead of durability or permanence. In the 1960s and 1970s, disposables grew ubiquitous. Fast-food restaurants like McDonald's used copious amounts of disposable containers, cups, plates, utensils, napkins, and other things. Soon companies marketed an array of disposable items for home use: paper plates, cups, napkins, towels, and tablecloths; plastic drinking straws and utensils; styrofoam cups; personal products; medical and cleaning gloves; dental floss; diapers; plastic bags and garbage bags; and on and on. No wonder that even clothing and furniture seemed disposable, too.[40] Each week, Amazon.com's warehouses send millions of used and returned items, often still in their packaging, to recycling centers, landfills, and incinerators because it is more profitable to dispose of items than to let them to take up space on warehouse shelves.[41]

Cities are running out of room to dump discarded consumer products. On the one hand, the weight of trash and garbage produced by the average American or European family has decreased over the past seventy-five years. People no longer throw away heavy items like glass or ash from coal-burning furnaces. Yet the sheer volume of waste fills landfills to capacity and every city must at some point find a new place to dump its waste. They do not find that easy to accomplish, of course, due to resistance from neighbors of suitable sites, who rarely welcome landfills next door. Floods, erosion, and other natural events often mean that what has been buried in old landfills does not stay buried.[42]

Age of plastic

The post-1970 era is the Age of Plastic. Cheap, durable, easy to form, taking any color from transparent to black, pliable or hard as needed, and versatile almost beyond imagining, plastics found thousands of everyday uses. Plastics' very durability and resistance to decay proves to be a huge environmental problem. Since they generally cannot be reused or repaired, most people discard or recycle plastics. As poorer nations with rudimentary or absent waste disposal services make and use plastics, environmental problems escalated spectacularly after 2000, especially in oceans. Vast "garbage patches" larger than most countries float in the center of the North Pacific, North Atlantic, South Pacific, South Atlantic, and Indian Oceans. On sea lanes floats an endless stream of plastic garbage. A 2019 expedition to the bottom of the Mariana Trench, the deepest place on earth, observed a plastic bag there. In addition, plastics pose dangers to human health. They contain compounds that interact with the endocrine system. They mimic hormones and disrupt sexual growth, development, and function, or promote cancer growth.

Plastic remains in the environment for millennia. Plastic bags in water can look edible and can entangle, suffocate, or clog the digestive tract of birds and sea animals. Plastic will break into smaller pieces rather than decompose. As plastic breaks down into tiny particles, water creatures at the bottom of the food chain ingest them and the particles travel up the food chain, concentrating as they go. Artificial microfibers also enter the food chain by means of laundry water. The Age of Plastics is choking on plastic waste.[43]

Landscapes of late consumer capitalism

With globalization of industrial and consumer capitalism came globalization of urban sprawl. In a single century, urban dwellers went from making up ten to 50 percent of world population.[44] Populations grew, migrations from rural areas quickened, incomes rose, and metropolitan areas expanded across neighboring landscapes. Each region has its own specific form of urban sprawl. Neither European nor Asian sprawl have the low densities of American sprawl. In Africa, where income is lower, cities have particularly high densities. Density corresponds to ways people move and transport goods and supplies. They concentrate closer together where they commonly walk or bike. Good public transportation or widespread ownership of motor vehicles encourages people to live farther away. Car ownership creates demand for road construction, which takes up land and lowers density further. Typical of all sprawling cities is unplanned, inefficient, "low-density, single-use, scattered, leapfrog development" with proliferating commercial strips. In contrast, as a symptom of economic and social problems, Russia's cities have shrunk since 1991—inverse sprawl.[45]

Sprawl's environmental effects are legion. At its most basic level, the spread of buildings and roads extends the area of impervious cover, which causes flooding and stops precipitation from percolating into aquifers. Water running quickly off streets flushes debris, trash, plastic, and oil left by motor vehicles into streams, rivers, and bays. Public services and infrastructure tend to fall behind in cities with corrupt, ineffective, or underfunded governments. Millions lack adequate clean water, sewers, roads, or waste disposal. Even if authorities construct waterworks and sanitary sewers, they usually lack funds to treat water or sewage. By their very nature, sprawling cities release more contaminants. Energy use for heating and cooling rises as density falls. Urban areas emit about 78 percent of the world's carbon, use 60 percent of its residential water, and burn 76 percent of wood used for industrial purposes. Pollutants in the air damage health and stunt vegetative growth. Rapid urban expansion also requires tremendous amounts of wood, cement, coal, steel, and other metals, each of which has its own environmental cost in terms of natural resource extraction, processing, and transportation. Finally, cities absorb heat during the day and radiate it into the atmosphere at night, which changes local weather patterns and increases demand for energy-hungry air conditioning.[46]

Energy: coal

Energy consumption rose with consumer capitalism's growth. Fossil carbon produces most of the world's energy, with cheap and widely available coal as the chief source of energy. Europe and North America continue to burn abundant quantities, but industrializing India and China have far surpassed them. Large, accessible coal deposits in these Asian giants supply most energy for industry and households. China now produces almost half the world's coal. The steel industries of both nations rely on coal, although the poor quality of India's coal forces the country to import better coal for steelmaking. Without alternative local fuel, India's powerplants burn its soft, smoky, sooty coal.

Postwar growth required more mining, the oldest and most environmentally destructive extractive industry. Despite America's voracious appetite for oil and gas energy, U.S. demand for coal energy surged in the 1950s and 1960s. The oil crises of the 1970s brought renewed attention to coal as an energy source to replace undependable foreign suppliers. The rolling green mountains of West Virginia and Kentucky would pay a terrible price. Coal mining had come to these states with the arrival of railroads in the 1880s, which made possible exploitation in remote regions of the same high-quality coal seams that ran through neighboring Pennsylvania. After 1945, mining companies increasingly turned to strip mining, which was cheaper and required fewer (and non-union) workers than traditional underground mines. In the 1970s and 1980s, development of huge, twenty-story high dragline excavators allowed the removal of the tops of mountains to get at the coal underneath, dumping the so-called overburden into valleys. Dust clouds and polluted watersheds plague local communities. When mining exhausts the coal, barren flats remain where green mountains once stood.[47] In the 1970s, American coal companies opened new regions to mining and developed cheaper but more environmentally destructive methods of strip-mining. Declining rail freight rates along with regulations on sulfur emissions promoted the development of thick, shallow beds of low-sulfur, low-ash coal in the remote Powder River Valley of Wyoming and Montana. By the early 2000s, gigantic excavating machinery allowed Wyoming's mere thirty open-pit mines, including the world's largest coal mine, to yield well over a million tons of coal every day, almost half a billion tons yearly.[48]

Strip mining is also common elsewhere in the world. In the 1970s, in Germany, the multinational energy company RWE bought the ancient Hambach Forest near Cologne in a region where open-pit coalmining had been common for a century. RWE cut down the forest and commenced digging up the soft brown lignite coal underneath. Since the 1990s, a gigantic bucket excavator, the tallest terrestrial vehicle on earth, has been biting deeply into the earth. Some of the overburden raises an artificial mountain, but much goes to refill nearby strip mines that have exhausted their coal. The excavator has dug down hundreds of feet below sea level. Pumping groundwater to keep the mine dry has lowered the region's water table and dried up streams and springs. Immense clouds of fine dust damage the health of townspeople in the area. Nearby powerplants burn Hambach's coal and send its toxic mercury content into the air. The notorious destruction of an ancient forest has made Hambach the best-known German open-pit lignite mine, but there are scores of others, particularly in former East Germany.

Environmental impacts of coal mining in China and India mirror those of Europe and North America, but their far larger and denser populations have magnified them tremendously. Vast open-pit coal mines operate in Inner Mongolia in western China and eastern and southeastern India and affect the health and safety of workers and the adjacent society, economy, and culture. Everyone breathes hazardous fine dust from mining operations. Wastewater from coal washeries carry particles, oil, and grease into streams and rivers. Waterways fill with acid mine drainage as well as water laced with heavy metals and toxic substances, pumped from underground and open-pit mines. Mine spoil dumps impoverish the health of the land, covering land with mountains poor in organic matter and deficient in nitrogen, phosphorous, and potassium. India struggles to properly dispose of coal ash from its low-quality coal, often laden with toxic metals. High sulfur content in Indian coal causes acid rain and ground-level ozone. The air of Beijing and New Delhi often dims with smoke and is dangerous for many millions to breathe.[49]

Energy: petroleum

Oil energy, the lifeblood of consumer capitalism, had many attractions. Many nations developed oil to transition away from smoky, sooty coal. Others had little coal but significant oil reserves. The United States and nations around the North Sea hoped to lessen their dependence on oil from volatile foreign regions. As the world's nations made the energy transition from coal to petroleum and natural gas, global production of oil and gas surpassed coal around 1970.

Several factors have made increased oil production possible. Between 1945 and 1970, dozens of major new fields came into production in the Persian Gulf region, the United States, the Soviet Union, Libya, and Algeria. After the oil shocks of 1973 and 1979 lent them extra urgency, major fields opened in Alaska in the U.S., the oil sands of Canada, the Persian Gulf nations, the nations of the North Sea, Mexico, Libya, Russia and Kazakhstan, Nigeria, Brazil, and China. Old oil fields have yielded abundant oil and gas due to improved methods of hydraulic fracturing ("fracking"), of drilling in deep water, and of processing oil sands. Fracking allowed U.S. oil and gas production to soar in the twenty-first century and made it a net oil exporter for the first time since the early 1970s. Nevertheless, since 1980 energy returns per energy invested has declined as oil production exceeded the amount in newly discovered fields. Production will decline in the future.[50]

Oil production is environmentally perilous. The *Torrey Canyon* was a harbinger of disastrous spills to come. In 1989, the *Exxon Valdez* struck a reef in the Prince William Sound in Alaska in the ecologically the most catastrophic spill. Accidents with the *Atlantic Empress* in Trinidad and Tobago in 1979, the *ABT Summer* off the coast of Angola in 1991, the *Castillo del Beliver* in South Africa in 1983, and the disastrous *Amoco Cadiz* off the coast of Brittany, France, in 1978 all spilled far greater amounts of oil than the *Exxon Valdez*. In the 1990s, new tankers were required to have double hulls. The size and number of large spills fell greatly.[51]

Far worse ecologically have been disasters on oil-drilling platforms and oil fields. An explosion and fire on the BP *Deepwater Horizon* in 2010 extensively damaged the fisheries and ecology of the Gulf of Mexico and the American Gulf Coast. A disaster on the Mexican platform *Ixtoc I* in 1979 had a similar effect on the southwestern Gulf of Mexico. On land, the deliberate destruction of oil fields in Kuwait and Iraq during the First Gulf War of 1991 unleashed an environmental firestorm. Hundreds and thousands of oil spills of all sizes around the world have received little publicity. Less dramatic have been the innumerable pipeline and rail tanker spills and accidents around the globe, some of which have produced significant, if localized, damage to land and watercourses. Pipelines are vulnerable to sabotage, theft, leaks, and accidents, but statistics on spills are hard to come by, since not all countries or companies reliably report them.[52]

Petroleum energy causes many environmental problems. Despite considerable efforts to reduce pollution from vehicle exhausts, cities regularly suffer from smog and ozone. Cities of poorer nations grew rapidly without much planning. Cheap or poorly maintained vehicles congest the roads and cover the sky with a photochemical haze. Karachi, Pakistan; Delhi, India; Beijing, China; Lagos, Nigeria; and Los Angeles rank as the cities with the world's worst air quality due to engine exhaust. Mountains that surround Los Angeles, Mexico City, and Kabul moreover keep smog from dissipating. With smog, blowing dust, garbage incineration, home fires, and coal-fired industries, citizens of poor nations breathe some of the world's worst urban air. Millions annually die early deaths.[53]

Pollution, air, and climate

The spread of industrial capitalism outward from Europe, the United States, and Japan to the rest of the world, with consumer capitalism following close on its heels, has brought immense changes to societies, landscapes, and oceans everywhere. Increasingly, and perhaps most alarmingly, it also altered the atmosphere upon which all living things depend. People could ignore air pollution if it was confined to urban areas. But it also produced acid rain, ozone depletion, and global warming, which respect no boundaries.

Smoke from sulfur-laced coal affects lakes, rivers, and forests downwind. Burned sulfur becomes sulfur dioxide, which combines with moisture in the air to make sulfuric acid. Precipitation returns the acid to earth to damage or kill aquatic life and forests. Tall smokestacks lessened local pollution only to spread it farther away. Downwind nations like Canada, Norway, Sweden, Poland, Japan, and the Philippines resented and protested damage caused by others' powerplants.[54]

In the 1960s and 1970s, manufacturers put many products into convenient spray cans. The best nontoxic, chemically inert propellant were chlorofluorocarbons (CFCs) like freon, the chemical that cooled refrigerators and air conditioners. In the early 1970s, scientists discovered that all the CFCs ever released into the atmosphere must still be there. CFCs worked their way up to the stratosphere, where the sun's energy caused a reaction that depleted the layer of ozone that protects life from harmful ultraviolet radiation. Additionally, scientists found that chemicals such as artificial fertilizers also emitted molecules that thinned the ozone layer. In 1985, satellites detected a disturbingly large hole in the ozone over the Antarctic and verified these theories.[55]

At the same time, scientific evidence grew stronger that the global climate was warming. Carbon dioxide from burning fossil fuels did most of the heating, but methane and other greenhouse gases that humans released intensified the effect. Polar regions have heated fastest. Glaciers are melting and disappearing, affecting people who depend on them for a steady summer supply of water. The Arctic, Greenland, and Antarctic ice caps are melting and raising sea levels, threatening to swamp coastal cities and habitats. Melting permafrost releases methane long frozen in the ground. Droughts become drier, wet spells wetter, and hurricanes and other storms stronger and more severe. Greater numbers of insect pests survive warmer winters. Plants and animals must migrate away from warmer, drier conditions for which they are not adapted, or die out. Forests dry out and burn, releasing more carbon. A global climate catastrophe is underway.

Ocean water has absorbed approximately half of the carbon dioxide humans have released since the dawn of the Industrial Revolution, creating carbonic acid. Ocean water is now acidic enough to affect living creatures. Shellfish, coral, and many other organisms make shells or skeletons with calcium carbonate, which dissolves in acidic water. Along with warming ocean habitats, which force fish and sea mammals to move to cooler waters, ocean acidification threatens to disrupt the entire ocean ecosystem.[56]

By the early twenty-first century, signs of all these effects were visible. Many of them tracked the upper limit that scientists had predicted as recently as 1990. The trends are now accelerating like a snowball rolling down a slope that threatens to become an avalanche.[57]

To infinity, and beyond

In 2018, Jeff Bezos was named the world's richest person and the wealthiest of modern times. On July 5, 2021, he stepped down from his post as chief executive officer of Amazon.com to become executive chairman.

His company and such global counterparts as Chinese online retailer Alibaba had played leading roles in speeding and transforming the global economy. They make consumption easier and cheaper than ever and convey unimaginable amounts of goods from supplier to consumer. Amazon.com and a handful of gigantic multinational corporations, answerable only to shareholders, now dominate many areas of the global economy.

NINE

The Rise and Globalization of Environmentalism

A movement is born

Until the 1960s, the general American public knew little of the complex interplay of government, manufacturing, advertising, salesmanship, and consumption, far less of their full impact on the environment. Then in 1962, the nation's best-known and best-loved nature writer published a scathing indictment of the chemical industry. Serialized in the magazine *The New Yorker* and released as a book that summer, Rachel Carson's *Silent Spring* caused an immediate sensation. President John F. Kennedy asked his science advisors to investigate its claims, which their 1963 report generally verified. The chemical industry reacted with an outraged counter-campaign against both book and the author personally. Some conservatives accused Carson of wanting to bring American agricultural productivity down to the level of that of Communist countries. So widely was the book discussed, debated, praised, and vilified, that it marked an unmistakable turning point in the transformation of the older conservation movement into the modern American environmental movement.[1]

One decade later, environmentalism stepped onto the world stage with the publication of *Only One Earth: The Care and Maintenance of a Small Planet.* It had been written by British economist Barbara Ward and French biologist René Dubos at the request of Maurice Strong, Canadian secretary general of the first Earth Summit at Stockholm. The book cited *Silent Spring* with approval but put its arguments into a broader context and proposed quite different solutions. Ward insisted that efforts to address global environmental problems must take into account the poverty of the global south. Here was the first formulation of the notion of sustainable development, couched in terms of postcolonialism, but not of capitalism, which Ward never mentions. *Only One Earth* looked forward with much greater prescience to the global environmental movement than *Silent Spring* did.

In developed countries, the environmental movement challenged the environmental problems of consumer capitalism, just as conservation had critiqued the heedless resource hunger of industrial capitalism. It emerged from prosperous societies first in the United States, a decade or two later in Europe, and then on a smaller scale elsewhere around the globe. A fragmented movement with no common agenda, environmentalism embraces a shifting coalition of interest groups variously focused on wildlife, wilderness, pollution, dams, environmental justice, population, and much else. Just as conservation did not fundamentally challenge industrial capitalism, environmentalism is a response to the environmental problems of consumer capitalism that has offered no coherent critique or realistic alternative to it. Just as captive as everything else to consumer values, environmentalism offers fixes and amelioration of problems. Perhaps, however, this is all that we can hope for.

A fable for tomorrow

Carson belonged to the last generation of a century of Reformed Protestant dominance in American culture, which was weakening and would fade to marginality by the 1970s. She was born in 1907 to Scotch-Irish Presbyterians in working-class Springdale, Pennsylvania, about sixteen miles upstream from Pittsburgh. Granddaughter and niece of Presbyterian ministers, Carson bore a lifelong seriousness of purpose. From early childhood, in her house at the edge of town, she grew up close to the natural world and participated in nature study, an activity that Reformed Protestants thought had great moral value to children. Carson graduated from the Presbyterian-affiliated Pennsylvania College for Women and earned a master's degree in biology from Johns Hopkins University. Poor family finances, her father's death, and the Depression ended plans for doctoral study. Carson took a position writing publicity material for the predecessor agency to the U.S. Fish and Wildlife Service. In her free time, she wrote scientific stories for popular publications. The emerging science of the sea fascinated Carson. She wrote three bestsellers about the oceans. Her writing success allowed her to quit her job in 1952 and write full-time.[2]

In the late 1950s, Carson turned to a topic that raised her righteous ire. A decade earlier, Carson had read reports of unintended effects and victims of DDT and other new pesticides. With a glowing wartime reputation for stopping insect-borne epidemics and saving lives, the chemicals were reaching the public in growing amounts. Too many farmers and government bodies used them indiscriminately, without regard to the dangers they posed to ecological or human health. Carson's sea books had connected her to an international network of scientists, who now passed on to her the latest science regarding insecticides.

Silent Spring's main contribution was to highlight thoughtless overuse of pesticides, but it went further and deeper. Carson described the interlocking system of educational institutions, government agencies, and private businesses that created and promoted toxic chemicals concocted in corporate labs. Agricultural schools and colleges benefited from corporate donors and steered research toward chemicals with potential agricultural uses. In pursuit of its long-standing purpose of increasing yields and decreasing farming risks, the U.S. Department of Agriculture encouraged farmers to use chemical pesticides, herbicides, fungicides, and fertilizers. Chemical-industry salesmen on commission had a financial incentive to sell as many chemicals to as many farmers as possible. Corporations also used bright, cheerful advertisements to sell dangerous and sometimes deadly chemicals to control pests in homes and gardens.

The result, Carson argued, was heedless use and abuse of chemicals. Local and state governments seeking to control pests sprayed supposedly harmless DDT over inhabited areas. Farmers and many homeowners had little reason other than cost to refrain from or limit applying chemicals. Many ignored warnings about how much and how frequently to apply products. Farmers found that insects that previously had not presented problems now became pests, even as target pests died off, because pesticides also killed creatures that had held them in check. Moreover, after several years, pesticides culled susceptible insects and left the few with genetic resistance to survive and multiply. Farmers found they had to use greater quantities of chemicals to achieve the same results or switch to more-lethal alternatives.

Silent Spring described the damage. Overuse or accidents sickened and killed. Pesticides and herbicides disrupted ecosystems. Fish and birds ate poisoned insects and were poisoned themselves or suffered reproductive problems, and in turn poisoned their predators. With each step up the food chain, chemicals grew more concentrated and lethal. Humans ingested them through residue on produce. Poisoning was just one danger. Evidence suggested a link to cancer.

Carson concluded with the argument that chemicals applied sparingly and carefully had their place in controlling insects. She discussed non-chemical solutions using natural controls for pests. She urged researchers and scientists to turn their research toward solutions that worked with nature and away from brute-force methods that wreaked ecological havoc.

Silent Spring's title suggests a book about declining birdlife, but passages on dangers to human health, including, for the first time, cancer, alarmed the public. Industry spokesmen tried to dismiss readers as birdwatchers and nature lovers, too romantic and emotional to understand the complexities of science and economics. The book stimulated research into problems caused by agricultural chemicals, which added to arguments to restrict or ban them. The chemical industry and its allies, however, concocted and marketed new chemicals, which now exist in profuse abundance, are sold globally, and continue to cause the problems Carson decried in 1962.

Silent Spring caught conservationists by surprise. It barely touched traditional issues of conservation of waters, soil, and resources, or of preservation of nature or wildlife. Its central point about the ecological interconnectedness between humans and the natural world also had far broader implications than traditional conservationism. Yet *Silent Spring* hearkened back to the moral heart of George Perkins Marsh's *Man and Nature* and of the conservation movement of Theodore Roosevelt and Gifford Pinchot. Carson's book was essentially a sermon, complete with an "altar call" at the end offering a choice between ecological redemption and chemical damnation. Greed for profit, said *Silent Spring*, was the root of all environmental evils. It corrupted government, educational institutions, and scientific research. Carson's moral arguments evoked a strong reaction in part because Reformed Protestant influence in American culture and politics still resonated with the public, and in part because most leading figures in incipient American environmentalism shared her Presbyterian moral perspective.[3]

Silent Spring traced the problem of pesticides to the workings of the modern consumer capitalism, but Carson never phrased it that way. Silence on capitalism no doubt constituted a necessary strategy during the Cold War, but Carson gave no thought to alternatives to the system she described. She only exposed corruption in the system.[4] Horror at unintended consequences of pesticides prompted Congress to ban or tightly regulate all the substances that Carson named, but never did the government seek deep reform or dissolve the bonds between government agencies, academic research, and corporations. Likewise, the environmental movement continued to tackle capitalism's environmental problems from pollution to global warming but has treated capitalism itself like kryptonite.

Development and environment

Only seven years younger than Carson, Ward enjoyed advantages Carson might only envy. Ward was born in 1914 to upper-middle-class parents in York, England. Her Quaker father was a barrister married, unusually for the time, to a devout Catholic. Unlike Carson, a lapsed Presbyterian, Ward was a lifelong active Catholic who absorbed from her Quaker father abiding concern for justice and the poor. Educated at a convent school in Paris, at the Sorbonne, and in Germany, she studied philosophy, politics, and economics at Oxford University, where she involved herself in social activism and Catholic groups. After graduation in 1935, she was appointed university extension lecturer, traveled to several countries to study political conditions, and began writing for newspapers and magazines. She published well-regarded books on international politics and colonialism. Like Carson a gifted writer, her work attracted attention. By World War II, she had joined the staff of *The Economist* magazine. When the Ministry of Information tapped her as a spokesperson for the war effort in 1942, she flew to the U.S. and spoke with people ranging from the White House to local Catholic groups. She observed the Nuremberg trials after the war and published books arguing for European economic unity and cooperation.[5] The theme of international cooperation would run through her life's work.

In 1950 at age 36, Ward married an Australian who held various postwar leadership positions at the United Nations and in Australia and Britain related to economic development. This changed the direction of her life. His jobs took them to Australia and to countries emerging from colonialism that wanted to develop their economies, like India and Ghana, where they lived many years in the 1950s. Unlike the childless Carson,

Ward bore a son in 1956. Unfortunately, like Carson she was diagnosed with breast cancer, in 1957. Cancer killed both women, Carson in 1964 and Ward upon its recurrence in 1981.[6]

Ward's expertise in economics and experience with postcolonial economic development made her a prominent university lecturer and advisor to leaders, including Presidents Kennedy and Johnson. She was an influential force behind the scenes at the Catholic council Vatican II, which oriented the the primary focus of Church social teachings to world poverty as a moral and religious issue. In the 1960s, she took an interest in the environmental crisis, its relationship to poverty in the global south, and northern countries' responsibility toward poorer nations, themes that dominated her books of the 1960s and 1970s. After Stockholm, she was appointed first president of the International Institute for Environment and Development and continued actively writing and speaking until her death.

Ward's accomplishment was to link the ideas of environment and development, which to the global north and global south had been separate issues. In her non-Protestant background and focus on the connection of justice for the poor to environmentalism, she reflected changes in environmentalism as it globalized after 1970. However, she did not analyze even in Carson's formative way how consumer capitalism drove poverty, development, and environmental problems. As implied by the title *Only One Earth*, Ward's solutions relied on the nations of the world coming together in good will and understanding to solve common problems.[7]

Interwar conservation

The environmental movement first arose in the U.S., where consumer capitalism had advanced furthest. did not arise out of nothing but had many important precursors.

America conservation had not stalled in the conservative reaction after the Progressive Era. On the contrary, conservation was still much on the minds of interwar Americans. In the 1910s and 1920s, increased leisure and prosperity, widespread ownership of automobiles, and improving roads broadened the appeal of camping, hunting, and fishing. New consumer goods like recreational vehicles, campers, Johnson outboard motors, and bright Coleman kerosene lanterns enticed people outdoors. A postwar flood of surplus army rifles armed a generation of sportsmen. At the same time, though, prosperity stimulated rapid economic development. Would-be fishermen found their favorite fishing holes filled in or polluted. Hunters discovered how much American wildlife had been already hunted out by sportsmen, farmers, and market hunters. Weapons manufacturers worried that, without game to kill, no one would buy guns and ammunition. In response, conservation organizations proliferated in the 1920s and 1930s.[8] In response to road-building and development in remote areas for automobile access, the Wilderness Society was organized in 1937.[9]

Major Depression-era initiatives of President Franklin Roosevelt and Secretary of the Interior Harold Ickes (a Presbyterian from western Pennsylvania) complemented and strengthened these private efforts. Roosevelt organized the Civilian Conservation Corps to put unemployed young men to work building park facilities, fighting erosion, planting trees, and doing other conservation tasks for state and national agencies. The CCC taught young men about the work of conservation and created an enthusiastic postwar audience for government environmental initiatives. Roosevelt also oversaw extension of protection for migratory birds and a massive expansion of the national wildlife refuge system. Ickes pushed for four new national parks designed more to preserve wild ecosystems than promote tourism or recreation.[10]

Ecological crisis on the Great Plains in the early 1930s stirred international alarm about human destruction of the land. Intense drought and heat hit a region that farmers had only begun to plow a few decades before. Winds hit bare fields and whipped up dark, terrifying, and sometimes deadly dust storms. In response, Roosevelt and Ickes established the Soil Conservation Service. The Dust Bowl also stimulated ecological research. Botanist Paul Sears's 1935 classic *Deserts on the March* popularized ecological concepts while warning of the global dangers of the desertification that the Dust Bowl exemplified. In Australia, Africa, the Soviet Union, and elsewhere, the notorious Dust Bowl served as a warning.[11]

Postwar environmental thinking

The U.S. entered postwar prosperity vividly aware of conservation, ecology, and the natural world. Rarely has the American public so eagerly read nature writing as during the 1950s, including books by Edwin Way Teale, Sigurd Olson, and Loren Eiseley. Carson's award-winning *The Sea Around Us* of 1951 stayed on the *New York Times* bestseller list for an unequaled 86 weeks. Her *Under the Sea Wind* and *On the Edge of the Sea* were also bestsellers.

Ecological ideas like those in *Silent Spring* circulated in larger global conversations about interconnectedness and relationships. The war's shocking brutality and stunningly abrupt atomic end prompted leaders across the Western world to cooperate to prevent another such calamity. Economic and ecological crises seem to have prepared fertile grounds in which fascism and military aggression took root. To foster international peace and security, the Allies organized the United Nations in 1944. The Bretton Woods Agreement and the General Agreement on Tariffs and Trade laid the basis for international prosperity. The United Nations Educational, Scientific, and Cultural Organization sought to foster internationalism in science and the humanities.[12]

Concepts of interconnection, interrelationship, and systems grew popular in other fields. Ecosystem, coined in 1935 by English ecologist Arthur Tansley, grew influential in biology. Others applied these terms to machines, "feedback loops," and society. Cybernetics and computer systems developed in the late 1940s. By the 1960s, systems, systems analysis, and similar terms enjoyed a general vogue in academic, governmental, and business cultures of the 1960s.[13] The popular ideal of one world promoted use of the new word *environment*, in the sense of the natural world around us, which came into use among natural scientists and ecologists in 1948. Carson made it a key term in *Silent Spring:* "The most alarming of all man's assaults upon the environment is the contamination of the air, earth, rivers, and sea with dangerous and even lethal materials."[14] Systems and cybernetics informed the Gaia theory of James Lovelock and Lynn Margulis in the 1970s.[15]

Antisystems thinking spread at the same time. Technical and social systems implied control or the possibility of control. To many postwar thinkers, control seemed suspiciously totalitarian. Carson's indictment of the interconnections between the U.S. Department of Agriculture, agricultural research institutions, and chemical companies for purposes of profit echoed President Dwight D. Eisenhower's warning of the power of the "military-industrial complex." It also looked ahead to 1960s radicals, who would decry the military-corporate system that caused racism, poverty, and war. In both *Silent Spring* and later countercultural publications, natural systems were good and profit or government systems sinister.[16]

Silent Spring played on the postwar fear of nuclear weapons and nuclear apocalypse. Apocalyptic thinking pervaded two bestsellers of 1948, William Vogt's *Road to Survival* and Fairfield Osborn's *Our Plundered Planet.* To Vogt, the famines, shortages, and disease epidemics of the war and immediate postwar period offered a portent of the future. Overpopulation, he wrote, was exhausting soils and depleting resources. *Road to Survival* made the first connection between human activities and the global environment. Osborn, a biologist, saw parallels between the recent war and humankind's war on nature. *Our Plundered Planet* warned that humanity was extracting so much from the earth that it was "becoming for the first time a *large-scale geological force*,"[17] an early statement of the idea of an Anthropocene epoch.

The apocalyptic mood worsened in 1949, when the Soviet Union's successful test of an atomic bomb set off a nuclear arms race. The end of human civilization loomed large. Another, quieter apocalyptic scenario was taking shape. Nuclear weapons tests in remote locations rendered the test site radioactive but also sent a menagerie of radioactive particles into the upper air. Scientists discovered that they rained back to earth far sooner than anticipated. They entered the food chain and grew more concentrated as they traveled from plant to animal to human consumer. Human muscle, bones, teeth, and organs accumulated radioactive elements that could cause cancer. In 1963, concerned governments agreed to ban open-air testing of nuclear weapons. *Silent Spring* explicitly played up how chemicals, like radioactivity, were ubiquitous, silent, invisible, tasteless, and potentially deadly.[18]

Genesis of a political and cultural movement

When Kennedy asked his science advisors to evaluate the claims of *Silent Spring*, the environment became a political issue. Congress held hearings, at which Carson testified, and continued to hold hearings on environmental problems throughout the 1960s and 1970s. Secretary of Interior Stewart Udall showed how environmental issues shaped the views of government with publication of his *The Quiet Crisis* in 1963.[19] Democratic presidential candidate Lyndon Johnson proclaimed his goal of a pollution-free Great Society in 1964, which paved the way for such major environmental legislation as the Water Quality Act of 1965, the Clean Waters Act of 1966, and the Air Quality Act of 1967.

Environmental problems now made national headlines with a frequency that they never had before, a reflection less of an increase in environmental problems than of new public awareness of them. Headlines prompted government action. A deadly smoggy four-day temperature inversion in New York City in 1966 led to the Air Quality Act. The wreck of the *Torrey Canyon* and other incidents raised awareness of oil spills, so that the oil spill off Santa Barbara, California, and the fire on the Cuyahoga River near oil refineries in Cleveland, Ohio, in 1969 culminated in tightened regulations in the Water Pollution Control Act amendments of 1972. The oil crisis of 1973 prompted President Jimmy Carter in 1977 to submit a comprehensive energy plan to Congress and create the Department of Energy. In the mid-1970s, toxic waste dumps repeatedly made headlines, the most prominent of which was the horror of a school and residential neighborhood built over an old toxic waste dump in Love Canal in Niagara Falls, New York. Congress responded with the Comprehensive Environmental Response, Compensation, and Liability Act of 1980, which created the Superfund to clean up sites contaminated with hazardous substances.

Protests against dam-building played a large role in the rise of popular environmental activism. Protests stopped a proposed dam in Dinosaur National Monument in 1955 and two dams that would have flooded parts of the Grand Canyon in 1968. Popular movements against ecological and social costs of large dams later boosted environmental awareness and activism with the Narada Dam in India, the Three Gorges Dam in China, and big dams in the Soviet Union, Europe, Africa, and South America.[20]

Local organizations sprang up to oppose nuclear power plants. The first one successfully opposed a proposed plant in Bodega Bay, California, close to the active San Andreas fault, which was cancelled in 1964. Protest movements in the 1970s fought unsuccessfully against Seabrook Station in New Hampshire and the Shoreham plant in New York. After the notorious accident at the Three Mile Island nuclear power station in Pennsylvania in 1979, however, no new American plants would be proposed and built until after 2000. Construction of plants proceeded in Europe, where the anti-nuclear-power movement would lay the basis for Green political parties in the 1980s.[21]

With such strong popular environmental consciousness, membership in environmental organizations rose rapidly. The National Wildlife Federation, Audubon Society, and Sierra Club each grew by tens of thousands in the 1960s and hundreds of thousands in the 1970s. New conservation organizations sprang up as well. The Environmental Defense Fund was established in 1967. The international environmental organization Friends of the Earth was founded in 1969. The National Resources Defense Council was organized in 1970. In the same year, American expatriates in Canada organized Greenpeace. Radical members split off from Greenpeace to form the Sea Shepard Conservation Society in 1977 and from the Wilderness Society to create Earth First! in 1980.[22]

Silent Spring also helped shape the values of the countercultural movement of the 1960s and 1970s. A generation grown up in relative comfort and security absorbed the criticisms of corporate, consumerist, conformist, suburban American society of the 1940s and 1950s and sought alternatives that gave greater meaning to life and work. Carson heightened fears that modern consumer-capitalist society was destroying both nature and human health. The counterculture contrasted the "system" and "plastic" modern suburban life with "natural" things, including natural fibers, natural foods, natural childbirth, natural medicines, natural personal products, naturism (nudism), and nature religion as expressed in neopagan rituals and indigenous spirituality. These accompanied rejection of mass-produced consumer goods, popularity of back-to-the-land movements, and desire to return to nature. By 1970, a boom in camping, backpacking, and hiking had commenced among the young.[23]

The 1970s, the American environmental decade

In the United States, these environmental, political, social, and cultural streams came together on April 22, 1970. Prompted by Senator Gaylord

Nelson and his staff, millions of people themselves organized local demonstrations, lectures, teach-ins, and a host of other peaceful events to mark Earth Day. It was a truly extraordinary moment in a very charged and divided era, when people of all political persuasions gathered to discuss the environment. Surprised politicians reacted quickly. President Richard Nixon established the Environmental Protection Agency. Over the next decade, Congress passed a host of environmental legislation. In 1970, it created Amtrak to restore passenger rail service. The Occupational Health and Safety Act of 1970 protected workers from dangers to health in workplaces. Following the second oil crisis in 1979, Congress funded a Synthetic Fuels Corporation and the Solar Energy Research Institute.

The year 1970 was not just an American turning point but an international one as well. American cultural influence washed over postwar Europe and the world through many channels. Hundreds of thousands of American soldiers introduced American music to European, Japanese, and South Korean youth. The vigorous, prodigious outpouring of American art, music, literature, television shows, and movies, often in translation, as well as clothing styles like blue jeans, followed them across the Atlantic and Pacific. Young Europeans were also keenly aware of American Civil Rights, campus, and antiwar protests, which had their European counterparts.

Environmental concerns took root abroad as well, although delayed in Europe while it recovered economically from World War II. Translations of *Silent Spring* appeared in a dozen languages. Reaction varied widely from nation to nation, but *Silent Spring* was never a bestseller in Europe. The Swedish response to *Silent Spring* exceeded that of the United States. Sweden had some of Europe's oldest and most popular nature protection organizations as well as a pragmatic, managerial approach to problems that made quick response possible. Sweden established the first comprehensive environmental regulatory agency, the Environmental Protection Board, in 1967 and passed the first comprehensive environmental protection legislation, the Environmental Protection Law, in 1968. Before anyone else, Sweden acted broadly against persistent pesticides.[24] In Britain, farmers did not rely on chemicals as much as Americans. The British government had for a decade already been dealing with the issue and responded now without much public debate. Quiet agreements with industry banned the most dangerous pesticides in 1964. The government gave this voluntary system a statutory basis with the Food and Environment Protection Act of 1985 and the Control of Pesticides Regulations of 1986.[25] In the Netherlands, *Silent Spring* made a large impression on scientists and government officials. The cumbersome Dutch government regulatory bureaucracy moved slowly until, in 1966–1967, DDT caused raptor populations to crash and pesticides killed thousands of terns. Regulation became much more effective, but without much public debate.[26] The West German Bundestag did not pass a comprehensive environmental law until 1971 and environmentalism played a relatively small political role until the 1980s.[27] In Denmark, Italy, Spain, and France, the press noticed *Silent Spring* and the American debate but concluded that overuse of chemicals was an American problem, and that was the end of it.[28]

By 1970, the steady environmental drumbeat could not be ignored. In 1968, Sweden proposed a United Nations conference on the environment in Stockholm. The Council of Europe declared 1970 European Conservation Year. The influential German magazine *Der Spiegel* featured its first cover story on the environment. After the American Earth Day, the pace picked up. In the first major demonstration against nuclear power, thousands marched against the station in Fesselden, Alsace, in France, in 1971. East Germany and France founded ministries for the environment in 1972 (although the East German ministry lost influence as the economy began to decline in the late 1970s). In 1972, Sweden's proposed conference, the United Nations Conference on the Human Environment, finally met, the first global Earth summit. It brought skeptical nations from the global south together with industrial nations, with China alone among Communist nations attending. Southern nations demanded an end to poverty before environmental considerations. Chinese delegates found the summit enlightening. Chinese environmentalism dates to this summit. In preparation for the conference, Ward and Dubos published *Only One Earth*. The conference led to the creation that year of the United Nations Environmental Programme. In 1974, agronomist René Dumont ran for president in France as the world's first ecologist candidate.[29]

Much less hopeful than Ward's *Only One Earth*, the Club of Rome's *The Limits to Growth* of 1972 shocked the world. *The Limits to Growth* relied on computer programs developed at the Massachusetts Institute of Technology to explore the interplay of exponential growth with finite resources. A gradual meeting of the ideas of earth systems and computer systems made it possible. American ecologist H. T. Odum employed cybernetic concepts in ecology immediately after Norbert Wiener coined the term in 1948,[30] while the American military funded research in computerized weather prediction. The Club of Rome team applied these methods to the general postwar worry about resources for the future as well as Vogt's and Osborn's concerns about overpopulation (amplified by American biologist Paul Ehrlich's 1968 bestseller *The Population Bomb*). *The Limits to Growth* predicted economic and demographic collapse in the middle of the twenty-first century. The book was the talk of the industrialized world. Millions of copies were sold in over thirty languages. Free-market economists derided it for underestimating the magical power of markets to find alternatives to diminishing resources. It nevertheless gave a huge boost to environmentalism, especially when the 1973 oil crisis and attending shortages appeared to announce that limits had already arrived. Several recent reassessments have concluded that the world is following the book's predictions surprisingly closely.[31]

Rise of the Greens

By 1980, global awareness of environmental issues had grown tremendously, but only in the United States had environmentalism become a political force. West German panic in the 1970s over the apparent death of forests from acid rain sharpened and popularized interest in the problem of pollution but produced no movement. A few industrial disasters raised concern about toxic chemicals, just as the same issue was stirring heated debate in the U.S. An accident in 1976 at a chemical plant in Seveso, Italy, released a cloud of toxic dioxin. Firehoses putting out a fire at a chemical plant in Basel, Switzerland, in 1986 washed pesticides and chemicals into the Rhine and killed fish and contaminated water supplies downstream all the way to the sea. The disastrous chemical plant accident in 1984 in Bhopal, India, which killed over 3,500 and injured 150,000, also inspired worldwide outrage but had little permanent political impact.[32]

The driver for change in France and Germany was the antinuclear movement. The leftist generation of 1968 recognized in its popular protests a political movement with parallel aims and came aboard, bringing political energy, suspicion of capitalism, and progressive social attitudes. As in *Silent Spring*, fears of radiation and cancer prompted by nuclear weapons provided the motivating theme. The 1986 catastrophe at Chernobyl sowed radiation and fear across Europe and boosted the fortunes and visibility of political environmentalism. New Green parties followed the French and German lead in virtually every democracy between Finland and New Zealand.[33]

In the United States, Reformed Protestants shaped the environmental movement and, in contrast, roots in the former slave states of the South or in the Southern diaspora correlate strongly with indifference to environmentalism. In Europe as well, culture and religious history influenced strength and style of Green movements. The Protestant north has had the strongest environmental politics and laws. In France, where Protestants are a small minority, a disproportionate number of major Green thinkers and figures have had Reformed-Protestant origins. Non-Protestant French Green leaders were often veterans of 1968.[34] West German Lutheran clergy had been involved in environmental matters since the 1950s and responded more vigorously than others to *Silent Spring*.[35] In East Germany, Lutheran churches provided a politically protected space for environmental critics of the regime's industrial policies to meet and organize.[36] In Catholic Alpine Germany and Austria, a strong non-religious, non-moralistic environmental movement arose, whose achievements include Germany's first national park, the Bavarian Forest National Park in 1970.[37] Strong environmental sentiment also prevailed in Lutheran Scandinavia, strengthened by a tradition of outdoor recreation, and the Protestant populations of the Netherlands and Switzerland. Norway has unexpectedly emerged as a sort of environmental paragon, fostering the Deep Ecology movement, founded in 1973 by philosopher Arne Naess.[38]

The internationalization of environmentalism

Only One Earth was a harbinger of the expanding dimensions of environmentalism in the 1980s and 1990s. It deepened to take in a social justice component and broadened to cover much of the globe. It is surely no coincidence that consciousness of capitalism's global reach also grew during the same decades—demonstrations against globalization greeted every major international economic summit of the 1990s. Nevertheless, single-issue movements have continued to dominate global environmentalism, with no major comprehensive proposals to resolve consumer capitalism's environmental impacts.

The movement for environmental justice grew out of local activism during the Love Canal crisis of 1978, where sickened residents discovered that their homes and schools were built on a former toxic waste dump.[39] Four years later, North Carolina created a dump for PCB-contaminated soil in a poor, rural, mostly black county. Protest erupted, led by black Baptist church groups and civil rights organizations. Similar protests against pollution and dumps in poor and minority communities across the nation sprang up in the next decade or two. The stretch of petrochemical plants along the Mississippi River between Baton Rouge and New Orleans, Louisiana, was one particularly egregious example of the toxic burden borne by the poor and powerless.[40] Some protests led to remediation efforts, but often corporations used the courts to evade punishment or responsibility.[41]

Elsewhere around the world, similar issues arose in defense of the rights of indigenous and peasant peoples against development. The Chipko movement of India grew particularly famous, in which villagers united to prevent the cutting of forests to which they had subsistence rights, a recent replay of old tension between peasant rights and development or government management. The plight of indigenous forest dwellers in the Amazon or Borneo also raised outrage among European and North American environmentalists. However, experience shows that peoples who successfully defend their right to use forest resources do not always manage them in ways that Western environmentalists consider sustainable.[42]

Ward was an early proponent of the idea of justice for the poor postcolonial global south from the wealthy north, which grew stronger in the years after Stockholm. In the 1990s, debt-for-nature trade became a common solution. However, although some countries' debts were paid off or forgiven in exchange for establishment of parks and wildlife preserves, the need to police and pay for them rendered many of them weak. Also, environmentalists of the global north worried about the denial of rights to resources of local villages or indigenous peoples in affected areas. The solution no longer has many proponents. The notion of north–south justice faded as well. More recently, concerns have grown about the effect of global warming on the poor, who with their low level of consumption

Figure 15. Bruna Prado. Brazilians gather in defense of the Amazon and protest against deforestation and forest fires on August 23, 2019, in Rio de Janeiro, Brazil. Beginning in the 1980s, the environmental movement globalized. Political protests and local actions are no longer confined to western Europe and the United States. (Getty)

bear least responsibility for climate change. Here Ward's activism bore fruit in the development of Catholic environmental thinking, which culminated in Pope Francis's 2015 encyclical *Laudato Si': On Care for Our Common Home* and similar statements.[43]

A particularly perceptive chapter of *Only One Earth* described the shared environment of the air and sea (including an early warning about global warming).[44] The 1980s saw the first global environmental crises. The first critical issue resulted in an unusually successful outcome. Scientists had shown in the 1970s that chlorofluorocarbons in aerosol sprays, refrigerators, air conditioners, fire extinguishers, and other uses threatened the ozone layer. Denials by manufacturers of CFCs ended in 1985, when satellite data gave convincing evidence of a large, growing hole in the ozone over the Antarctic. International negotiations began quickly and led to the Montreal Protocol of 1987, which created the regulatory framework that led to a declining atmospheric concentration of CFCs.

Ward also warned about decline of tropical rainforests, which in the late 1970s and the 1980s stirred widespread alarm.[45] A radical environmentalist offshoot of Earth First!, the Rainforest Action Network, organized in 1985 and urged boycotts of companies whose actions contributed to tropical deforestation, with remarkable success during the organization's first decade.[46] Harvard biologist E. O. Wilson popularized the concept of biodiversity with his 1988 book *Biodiversity* and gave environmentalists a powerful new concept to defend undeveloped forests. Only when accelerating climate change in the twenty-first century pushed to the fore the value of rainforests for carbon storage did biodiversity fade somewhat from the conversation. However, reprieves for rainforests, such as the Amazon's, have often been temporary, especially when a change of government unleashes loggers, miners, ranchers, and other agents of deforestation.

Beginning in the Stockholm Conference in 1972, poorer nations have made clear the priority of economic development over environmental issues. This, of course, poses a major hurdle for proposals from citizens of wealthy northern nations to dismantle capitalism to save the environment. In 1983, the United Nations established a Commission on Environment and Development, known as the Brundtland Commission after its chair, former Norwegian prime minister Gro Harlem Brundtland. The commission's 1987 report, *Our Common Future*, popularized the phrase "sustainable development," which it defined as development to meet needs of the present without compromising needs of the future.

Successor Earth summits now convene every decade, in Rio de Janeiro, Brazil, in 1992, Johannesburg, South Africa, in 2002, and again in Rio in 2012, each featuring the theme of sustainable development. The U.S. has participated without enthusiasm, largely due to pro-corporate Republican administrations in office in 1972, 1982 (which blocked an Earth summit in Nairobi, Kenya), 1992, and 2002, and partly due to the growing influence of neoliberal sentiment within the Democratic Party, in office in 2012. Nevertheless, beginning with the first Rio summit, a raft of new international environmental agreements ensued on climate change, biological diversity, tropical timber, desertification, international traffic in hazardous substances, plant genetic resources, ocean fisheries, and renewable energy.[47]

Failure to address the warming climate

The biggest international failure in solving a global environmental issue has been failure to slow or reverse climate change. Environmentalists have been unable to overcome resistance from fossil-fuel and corporate interests and from states whose economies depend on them. By 1980, global warming already troubled scientists but not yet the public or politicians. In time, scientists acquired more evidence, computer models grew more sophisticated and accurate, and disturbing future scenarios emerged from a series of international scientific climate conferences. Scientists grew desperate to affect government policy to avert climate catastrophe in the first half of the twenty-first century.

In 1988, a year in the U.S. of record heat, severe drought, devastating forest fires, and a powerful hurricane, the American public and press began to attribute natural disasters to global warming. Bill McKibben's bestseller *The End of Nature* of 1989, the first book on climate change for a general audience, also provoked a conversation on climate.[48] British Prime Minister Margaret Thatcher, a former chemist who understood the science, called for action and greater research. American Republican President Ronald Reagan demanded an independent agency that Republicans hoped governments could influence politically. United Nations agencies responded with the Intergovernmental Panel on Climate Change. The very size and international composition of the IPCC lent its work tremendous credibility, even as inclusion of scientists from petroleum states like Saudi Arabia skewed its conclusions toward caution and conservativism. In 1990, the IPCC began releasing regular reports. Since then, cautious phrasing has evolved into more emphatic language for government action. Its worst-case predictions have tended to match reality disturbingly closely.[49]

The 1992 Rio convention on climate change led to the Kyoto Protocol in 1997, which committed signatory nations to reduce their production of greenhouse gases. However, the U.S. under Democratic President Bill Clinton failed to ratify it and his successor, Republican President George W. Bush, repudiated it altogether. A followup conference in Copenhagen in 2009 failed, but the ambitious successor agreement negotiated in Paris in 2015 included almost all the world's nations. Then, to widespread criticism and resistance, Republican President Donald Trump withdrew in

2020, only to have Democratic President Joe Biden rejoin the agreement in 2021. The Paris Agreement provided a yardstick by which nations can measure success or failure but did not impose firm commitments to cutting greenhouse gases.[50]

The climate change issue changed the nature of global environmentalism. It now overshadows most earlier concerns of conservation and pollution. Current priority lies in a transition away from fossil fuels to renewable energy as quickly as possible. By 2021, catastrophes loomed like something from an environmentalist's most apocalyptic scenario. Forest fires burned out of control in Siberia, the American west, Greece, and Turkey. The government of President Jair Bolsanaro allowed accelerated clearing of Brazil's Amazon rainforest. Record droughts seared some places while record torrential rains flooded others. Hurricanes, typhoons, and cyclones grow larger and dump more rain. The polar and Greenland icecaps are melting faster than researchers had forecast. Low-lying cities and islands faced rising oceans. Climate change threatens species with extinction. Derangement menaces ecosystems almost everywhere. No environmental crisis had ever approached climate change in its scope, urgency, and threat to human civilization.

The environmental challenge of corporate agriculture

Environmentalism instigated a backlash against the industrial-agriculture complex of chemical-dependent monocultures of genetically modified plants and factory-style production of animal products. American government policy, especially under the Republican administrations of Eisenhower, Nixon, and Reagan, had encouraged American farms to grow larger, while the Agriculture Department promoted mechanization, artificial fertilizers, and chemical pesticides, herbicides, and fungicides to make them more productive. The Green Revolution spread these trends throughout the global south. *Only One Earth* praised the ability of the Green Revolution to better feed the world, while cautioning against problems of chemicals, erosion, and salinization.[51]

In Europe, the founding of the European Common Market in 1957 promoted free trade in agriculture, while allowing each nation to protect its farmers from competition. Price guarantees protected small, inefficient farms. Europeans could protest with some truth that *Silent Spring* applied to the United States but not to them. In 1970, the European Community adopted the Mansholt Plan to make European farms more competitive and less dependent on expensive price supports. Expensive price supports did not disappear, however, so more-productive, larger farms create large agricultural surpluses even as Europeans eat more expensive food than Americans. France became the world's second largest food exporter, after the United States. The politically rigid, cumbersome European agricultural regulatory system has been forced to change since 1990 under the pressure of globalization, the eastward expansion of the European Union, and consumer and environmental groups. Yet agriculture continues to be highly regulated and very costly. In both Europe and the U.S., agribusiness and huge agricultural corporations now control almost all food products.[52]

In reaction, a movement developed for organic meat and produce. In the first half of the twentieth century, marginal precursor movements arose in Germany, England, and the United States and spread to neighboring countries. The German biodynamic system and English organic movement promoted health of the soil with manure and crop rotation to bolster fertility, often in a spiritual, mystical, or religious context. The English movement took inspiration from experience in British India. As promoted by Jerome Irvine Rodale (born Cohen) and others in publications like Rodale's journal *Prevention*, the movement in the U.S. concentrated on human health. *Silent Spring* firmly connected healthy ecosystems and human health in the public mind. Later in the 1960s, researchers linked common food additives and artificial sweeteners to cancer as well. The American organic movement emerged out of the counterculture among people who wanted to eat healthy food raised without chemicals and to escape the power of agribusiness and large food corporations.

In the 1980s and 1990s, the younger generation eating organic foods matured and prospered. Large food corporations saw a potential market. That, plus a reaction against introduction of genetically modified organisms, fed demand for production of organic foods. Organics appeared on the shelves of chain supermarkets and large retailers like Walmart. Consumers also recoiled at the overuse of antibiotics and hormones in the production of meat and dairy products. Since 2000, desire for a carbon-neutral agricultural system has prompted another growth in demand for organics, along with popularity of locally produced foodstuffs. Furthermore, in the 2010s, widely reported scientific studies described dramatic declines of insect species due to pesticides, notably vital pollinators like bees. The havoc that chemically dependent monocultures caused had worsened since *Silent Spring*.

Consumer capitalism soon captured the smalltime organic food system. As mainstream shoppers bought organic food in the 1980s, governments and private organizations certified that producers or marketers had not misrepresented conventional foods as more-expensive organic. Rules were necessary for trust in the organic system, but bureaucratization also allowed large farmers and corporations into the organic marketplace. In both the U.S. and Europe, health-food stores and organic farms began to resemble their industrial and corporate counterparts. Careful farmers using traditional methods disappeared. Agribusiness and corporations took over. Multinational corporations bought up small organic brands. A social movement to reform the way society produces and eats its food transformed into another system of mass production for the consumer market.

The problem of securing the quality of meat raised industrially has proven troublesome. Europeans regarded meats with hormones and antibiotics as unhealthy and attempted to restrict imports from the United States, where regulations tended to favor producers' interests. In 1997, researchers showed that "mad-cow disease," bovine spongiform encephalopathy, could cause an incurable brain disease in humans who ate infected beef. The disease affected British beef fed with tainted tissue from sheep (obviously, not natural food for cows). British beef exports stopped and affected herds were destroyed. At the same time, horror stories circulated in the press about common inhumane conditions in concentrated animal feeding operations. These events fed suspicion of corporate food and raised demand for traditionally raised meats.

The Green Revolution in Latin America, Africa, and Asia forced economically inefficient small farmers to move to growing cities. Americans and Europeans bewailed the loss of ancient locally adapted agricultural techniques, seed varieties, and rural cultures. American and European activists also attacked large American and European corporations that took genetic material and plant varieties from poor countries without compensation and sold it back to them incorporated in commercial seed.[53]

International demand for organic, less environmentally harmful, less exploitative, and less polluting agriculture grew, first in Europe and the U.S., then in Japan, Australia, and New Zealand. Banana, coffee, and tea plantations that export to those countries adopted organic techniques as market demand shifted. Sometimes, consumer demand can change the system.

Corporations and the environment

While consumer capitalism is the engine that drives environmental problems, it is not intrinsically hostile to a greener world. The response of corporations to environmental problems varies considerably. Certainly, powerful factors are at work that slow or block environmental progress. The duty to uphold shareholder value forces corporations to keep their eye on quarterly or annual reports and rarely on long-term goals. Then, too, the inertia and conservatism built into large organizations prevent them from responding nimbly and creatively. Large investments in existing material assets also inhibit rapid change. The global fossil-fuels industry, for example, has trillions of dollars invested in oil and coal reserves, production equipment, pipelines, refineries and plants, and service stations, and cannot easily divest or abandon them.

Where owners or chief executive officers have strong conservative or libertarian convictions, management may lobby against or obstruct government regulations, or delay them through court proceedings. Moreover, the nature of the American tradition of oil wildcatting, and perhaps the nature of extractive industries themselves, means that owners and executives of oil and gas companies, like Charles Koch, for example, strongly oppose government regulation. Even in relatively green nations like Germany, environmental regulations can run into obstacles. There, a car-loving culture and the power of German automakers prevented the government from phasing out leaded gasoline, requiring catalytic converters, or imposing speed limits until 1996.[54]

Yet corporations can also embrace environmental goals if they can profit by it, sometimes after regulations force them to change their methods or products. The necessity of changing energy sources to forestall oncoming disasters of runaway climate change has spurred a great deal of innovation, particularly in California. Entrepreneurs like Elon Musk have enlisted wealthy Silicon Valley investors to develop electric vehicles and better batteries. Musk's success has forced other automakers to also produce electric vehicles. Another entrepreneur backed by Silicon Valley investors, JoeBen Bevirt, founded Joby Aviation to design an electric vertical take-off and landing commuter aircraft as an air-taxi service comparable to Uber and Lyft and plans to offer passenger service in 2024. As solar and wind energy grows more common, economies of scale make it a cheaper, cleaner alternative to fossil-fuel energy that is attractive to power companies and investors. Prompted and partially subsidized by government, such companies will lead the way to an ecologically better future.

Government laws, regulations, taxes, subsidies, and incentives direct and spur innovation. Environmental governance emerges out of an interplay between the forces of corporate interest, government objectives, and public opinion as shaped by media. Media or scientific reports raise no environmental alarms. The public only demands action after particularly notable crises that concentrate attention on problems. Democratic governments responsible to public opinion prove a great deal more responsive to environmental problems than centralized, authoritarian governments. Hence, European socialist nations before 1991 tended to have horrible unaddressed environmental problems. China steers a middle path between Soviet environmental incompetence and Western democratic environmental governance. Its large size and huge population make it less governable by fiat. Its Communist government has had to delegate significant power to local authorities but does respond when too many complaints reach it.[55]

The three-way relationship between corporations, public, and government is not equal. Companies have great sums to spend on public relations, lobbying, and political donations. For example, Koch's effective network of think-tanks, foundations, institutes, and lobbying groups shows how the wealthy shape public opinion and corrupt the democratic process.[56]

Companies may deflect responsibility away from themselves and portray environmental problems as products of consumer choice. If consumers blame themselves for litter and waste, for example ("We have met the enemy and he is us," in cartoonist Walt Kelly's famous aphorism),[57] they will ignore corporations that profit from disposables. The American Can Company, Coca-Cola, and other businesses developed the

"Keep America Beautiful" advertising campaign for this purpose in the 1960s. Other advertising campaigns urged people to recycle plastic waste, despite the limited market for recycled plastics, or to reduce their ecological "footprint" through flying or driving less or eating less beef. Fossil-fuel companies have run advertisements casting doubt on clean energy solutions and touting "clean natural gas" or "clean coal." They have lobbied legislatures to tax renewable energy or electric vehicles or to eliminate government subsidies or tax breaks for competing energy sources.[58]

Corporations have many tactics to delay, forestall, or undermine environmental regulations. They use networks of sympathetic scientists to disseminate doubt about the science of environmental issues which have delayed government action on acid rain, the ozone layer, and global warming.[59] They might participate in investigatory hearings, give assurance of changing practices, then not follow through with substantive changes. They can restrict access to information or industrial sites to prevent government action. They promise that the free market would promote development of technology to solve the problem. They can tie up regulations in the courts and hope for a more sympathetic government to take office. Finally, corporations protest loudly that government regulation hurts the economy.[60]

Capitalism and environmental movements

Global environmental movements do have their victories. The water and air of many nations have become much cleaner. Certain species have been rescued from the brink of extinction. Regulatory regimes are in place to prevent certain problems and abuses. Still, however, the world lurches from one environmental crisis to another.

Rachel Carson's *Silent Spring* criticized the corrupting effect of the profit motive. The solutions she proposed, however, did nothing to address the tight nexus between business, academia, and government that promoted irresponsible use of pesticides. *Silent Spring* opens with the tale of a fictitious incident in a small rural town, the very sort of place that growing agribusiness even then was dooming to depopulation and poverty. With her predilections toward moralism from her religious upbringing, Carson regarded use of chemical pesticides and herbicides as a moral choice rather than product of a consumer-capitalist society. Similarly, the many American environmentalists with Reformed Protestant backgrounds have blamed greed and self-interest for destruction of wild places, with not a word about capitalism itself.

The European Green movement emerged in a union of leftist radicalism and antinuclear protests. In its early days, it voiced strong critiques of capitalism. After the 1990s, as the movement matured and entered governments, anti-capitalist rhetoric mellowed. Some early Green leaders, like France's Brice Lalonde and Germany's Joschka Fischer, moved to the right as they aged. Lalonde turned from opponent to proponent of nuclear power. Fischer retired from politics to lobby for former nuclear-power proponent Siemens, luxury-carmaker BMW, and energy and coal-mining giant RWE.[61]

Environmentalists have never spoken with one voice. Many attack consumption as an individual moral choice, as if a choice to buy an electric car, recycle, and eat organic foods would themselves save the earth. Corporations favor this perspective since it places responsibility for environmental problems on consumers and not themselves. A few American environmentalists on the right, among them Dave Foreman and Michael Shellenberger, argue that free-market mechanisms will solve environmental problems. On the far left, a handful of anarchists, notably Murray Bookchin, reject the corporate-capitalist system and imagine small, ecologically responsible communities living in harmony with nature. American leftist environmentalism never enjoyed large numbers of adherents. Biologist Barry Commoner was something of an exception. Still, Commoner's 1971 bestseller, *The Closing Circle*, has been more cited for its "laws of ecology" than for its criticism of capitalism and advocacy of socialism. Some today, most prominently Jason Moore, continue to advocate socialism and assign responsibility for the environmental crisis to capitalism, excusing Soviet socialism's dismal environmental record by insisting that it wasn't true socialism.

For all its strengths, Ward's *Only One Earth* offered no firm solutions to the problems it described. Her hopes for concerted worldwide action in the face of looming crisis have failed to produce results at a half-dozen Earth summits. The moral duty toward the poor has not prevented the gap between the wealthiest and poorest nations from widening instead of closing.

What, then, can be done?

Conclusion
Profit — Capitalism and Environment

Does it profit us when someone else makes a profit? The answer is unclear. In the pre-modern Christian West, monetary profit entailed a moral calculus. People observed that, in agrarian communities, one person gained only if someone else lost. In moral terms, greed and avarice must push somebody into poverty to make another person rich. To balance the moral scales, a wealthy person must give back to society so that others might profit from their good fortune. A similar calculus inspired Andrew Carnegie and other capitalists to give their wealth for the greater social good. John D. Rockefeller did the same, but he also realized, as did other corporations, that charitable giving enhanced Standard Oil's public image. Returning one's wealth to society still today motivates some of the superwealthy, although far from all of them.

Modern economic thinking recognizes that the economy is not a zero-sum game and that incomes of both poor and rich can rise together, as they have, if unequally, since industrialization and, more equally, from 1945 to the 1970s. The postwar moral calculus focused on ensuring that the gap between rich and poor did not widen. Attitudes shifted again in the 1970s and 1980s as neoliberal orthodoxy swept corporate leadership. Policies of deregulation, lower taxes on the wealthy and corporations, and focus on shareholder value rest on assumptions that prosperity for some translates to prosperity for all. The assumption has turned out to be false, as income for the richest rises like a rocket, middle-class income stagnates, and the income for the bottom quintile even falls. A more all-encompassing form of corporate propaganda called public relations replaced advertising departments and absorbed the duties of charitable giving. Charity became another business expense, part of a public relations campaign and justifiable only to the degree it increased profits. Corporate charitable giving has fallen by half since 1980. In corporate boardrooms, moral calculus has given way to profit calculus. In society at large, where consumer values are displacing traditional religious ones, self-interested greed now walks unashamed.[1]

Do the profits of capitalism profit the rest of us, then? Capitalism's defenders can point to pre-pandemic decreasing global poverty[2] (if not decreasing inequality) as evidence it does. Certainly, you, the average reader of this book, will have benefited greatly from consumer capitalism. Chances are, you live in snug, well-built housing. You choose from an abundance of clean clothing to dress every morning. Your streets, water, and food are almost always sanitary and clean. Your health care is almost always safe and effective and involves no leeches, laudanum, or magical rituals. You likely know no one whom a work-related accident killed or maimed. You wear inexpensive, well-fitted clothing. You eat abundant, cheap food. Most of us can climb into a heated or cooled vehicle or convenient transport and comfortably and cheaply travel at our convenience anywhere from the grocery store to the far end of the continent. In the space of a day or two, you can travel remarkably cheaply almost anywhere on the planet at 37,000 feet in a cushioned seat breathing pressurized room-temperature air. You can open up a portable electronic device and watch events as they are happening many thousands of miles away, or read a book or poem, or send or receive messages instantly, or entertain yourself with an immense selection of movies or music, or write a book like this one.

Unfortunately, as this book has shown, we profit and have always profited at nature's expense. Here the moral calculus of George Perkins Marsh and William Stanley Jevons applies. The earth can no longer bear the burden of supplying raw materials and absorbing the waste to make the consumer capitalist machine run. The earth is our dump from the bottom of the ocean to low orbit and from the equator to the poles. A cascade of environmental disasters of every sort threaten to overwhelm human civilization. The planet is not disposable, to be discarded when soiled. Despite fantasies of certain billionaires, no unspoiled planet exists to which we can flee. No alien Star Gate floating through space has appeared to guide us to a happier future.

How did we get here? Modern consumer capitalism is the product of history, but it was not inevitable. History is too full of contingency and chance for that. Had the Genoese prevailed at the War of Chioggia, had Columbus drowned when his ship sank off the Portuguese coast, had James Watt not been tapped by Glasgow University to repair astronomical equipment, had Stanford University not been connected to ARPANET, or had any of a thousand million other human events gone otherwise, the world would be a different place today. Capitalism would nonetheless be here in some form. It is rooted in human nature and human history. These deep roots, some of which go back to our remotest ancestors, make capitalism resilient and adaptable to time and circumstances, so that the capitalism of one time and place is not that of another. These roots also make it extraordinarily difficult to replace. Throughout history, those who would have killed or constrained it have found that it always seems to grow back. To paraphrase the Bible, capitalism we have always with us.

This book traces the environmental impacts of capitalism's germination and growth through human history. Since the dawn of humanity, the success of *Homo sapiens* allowed them to increase and multiply. Population growth pressed people to develop new ways to feed and care for each other. Tools allowed more efficient harvesting of plants and animals but left fewer of them. When resources grew short, groups might migrate to a less exploited region. Or they might aggressively slaughter, enslave, or displace weaker groups and take their productive land. Some groups found outcroppings of valuable rocks and minerals and mined them to trade for items they did not have to produce themselves. When no other option was available, human ingenuity turned to exploiting resources more intensively with agriculture or herding, and traded surplus for material desirable for ritual, adornment, or use.

The cycle repeated for hundreds of millennia. Population grew denser, promoted more sophisticated trade, and developed new ways to exploit resources more efficiently. Surpluses became capital that could be invested in trade, production, or displays of wealth and power. Water, wind, wood, and domesticated animals provided energy for processing and manufacturing goods. Writing allowed faster, more reliable communication and recordkeeping, especially after the development of alphabets. People around the world figured out how to turn seeds into bread and beer; clay into pots; plant, animal, and insect fibers into clothing and rope; rocks into metal objects; plants, insects, minerals, and shellfish into dyes; and much more. All became objects of desire and trade. To carry them farther, boats grew larger, more efficient, and more reliable, but also became suitable as instruments of war and colonization. Money simplified and encouraged trade. Arabic numerals made accurate bookkeeping possible. Trade partnerships pooling capital evolved into permanent trade houses, which, with proper legal support, laid the basis for corporations, with even larger pools of capital. Printing allowed easier, cheaper dissemination of information.

At last, in the eighteenth century, wage labor began to replace unfree labor, even if slavery and servitude survived a while longer. The energy and power of machines burning fossil fuels opened tremendous possibilities for production and transportation. Manufacturing became cheap and standardized. Transportation sped up on land and sea. Versatile, adaptable electric power put energy into homes, workplaces, vehicles, and, today, into small devices in our pockets and purses. Communication flew at the speed of light over wires and radio waves. Corporations took advantage of print and electronic media to advertise their products and entice or manipulate people into consuming the great flood of products they manufactured. Bankers found new ways to loan people money for buying things. Consumer capitalism had arrived and begun its victorious march around the world.

Consumer capitalism's inherent drive to entice people to consume ever more has drained the earth of its fertility, resources, and biological diversity, and threatens a dire future from pollution, extinctions, and global warming. Just as conservation never challenged industrial capitalism but only offered to fix its problems, environmentalism accepts consumer capitalism while challenging its excesses. Rachel Carson and Barbara Ward diagnosed the problems from moral high ground but failed to propose effective remedies. Environmentalism has only mitigated some of capitalism's worst abuses. It may yet rein in climate change, but the fossil-fuel industry's propaganda and influence on governments has forced delays that leave us at the brink of disaster. As the crisis intensifies, political will may finally be at the point where necessary progress can be made.

Yet, however much it might spare wildlife and clean the land, water, and air, we stop the machinery of consumer capitalism at our peril. Whenever capitalism slows or verges on seizing up altogether, people lose jobs, go hungry, lose their homes, and look for scapegoats and extremist solutions. The Great Depression and the wave of fascism and war it provoked serve us as a warning. Disaster threatened again when deregulated banking and finance nearly crashed the global economy in 2008. Recovery was still incomplete when the COVID-19 pandemic hit in 2020. The pandemic cleaned urban skies around the world but also disrupted economies and threw people out of work.[3]

Critics of capitalism regard it as a profit-driven system responsible for environmental destruction, global warming, racism, colonialism and imperialism, oppression of workers, and a whole host of other sins. This may be true, although most of these problems have existed all over the world in various forms at most times since the development of civilization, or longer. We should not expect that after the end of capitalism peace will rule the planets and love will steer the stars.

Realistic alternatives are scarce, unfortunately. Twentieth-century state socialism stands thoroughly discredited from the point of view of either social or environmental justice. Murray Bookchin envisioned small democratic communities living close to nature. Bookchin is extremely vague on how we get from where we are to his little communities, and vaguer still on how to keep them small and independent. Something like Bookchin's vision was attempted during the late 1960s and early 1970s, when subsistence communes were briefly popular in California, Oregon, Vermont, and elsewhere. Their members aspired to escape the evils of capitalist society, find personal fulfillment, and live harmoniously with nature. Almost none of those communes lasted long. Hard reality and disillusioned romantic notions soon sent almost everyone on them back to consumer capitalist society. When I lived in San Francisco in the late 1970s, I knew several people who had lived on communes but had come back to the city. I recall hearing of one group that went to India to live on a communal subsistence farm. They gave up and returned to the West after monkeys stole the fruit and cobras took up residence under the porch.

The temptations of consumer capitalism nearly always prove irresistible. As a boy in the late 1920s and 1930s, my father lived a life barely touched by consumer capitalism. He grew up on a farm. He did live in a Sears prefabricated house, his father farmed for the market with a tractor, and the family drove to town in their Ford every Saturday, but there were few other contacts with the wider capitalist world of consumption. At Christmas, for example, he might receive a book under the tree. His family had no indoor plumbing or electricity, bathed once a week all in the same bathwater (pumped by hand from a pump in the yard and heated on the wood-burning stove), and worked outside all summer in the humid southeast Kansas heat. Farm accidents killed his cousin and his stepmother's first husband. My father eagerly left that life behind and never wanted to go back.

What, then, can be done?

Like a shark, consumer capitalism must keep moving to live. If it slows or stops, dislocation, unemployment, unrest, and wars plague the earth. We know the shark needs to take a different path. The only requirement to keep consumer capitalism running is to keep as much money flowing into as many pockets as possible. The challenge may be to do so with as little demand for resources as possible.

We may already be moving in that direction, however tentatively. In January 2016, the furniture giant Ikea's chief sustainability officer Steve Howard declared the approach of "peak stuff," when people would not need any more furniture. He noted that people already bought less gasoline, less beef, and less sugar, and that furniture sales had also begun to plateau.[4] We are making a transition from buying stuff to buying other things. Over the past half century, people have begun to buy experiences rather than stuff. They go on cruises, travel to foreign countries, climb mountains, scuba-dive among coral reefs, and go on other "adventures" (in quotation marks to acknowledge how canned and safe most of them are). People eat more often at restaurants or drink at bars. Entertainment that demands very little of the earth can be found on every smartphone, tablet, or home computer. Consumers can spend money—the vital activity upon which consumer capitalism depends—on video games, movies, television series, and many other things, without buying a physical object. Unfortunately, this sends little money to the global south. Playing Minecraft instead of, say, buying a T-shirt, puts no money in the hands of a poor factory worker in the Dominican Republic. Nevertheless, buying less stuff seems critical to securing the future of us and the earth.

So, like it or not, we are captives on this accelerating merry-go-round of consumer capitalism. Adopting renewable energy to slow global warming seems to be the most important first step. Breaking up big corporations with their tremendous economic and political power and setting curbs to their propaganda would surely be essential to any comprehensive solution. We can only hope it will be possible. The collapse of civilization predicted fifty years ago by *The Limits to Growth* no longer seems like a distant threat. The clock is ticking very loudly.

Notes

Introduction

1 Mariana Mazzucato, *The Entrepreneurial State: Debunking Public vs. Private Sector Myths*, rev. edn. (New York: PublicAffairs, 2015), 93–120; Elizabeth Jardim, *From Smart to Senseless: The Global Impact of 10 Years of Smartphones* (Washington, D.C.: Greenpeace, 2017); Jason C. K. Lee and Zongguo Wen, "Rare Earths from Mines to Metals: Comparing Environmental Impacts from China's Main Production Pathways," *Journal of Industrial Ecology* 21(5) (October 2017): 1277–1290; Amnesty International, *"This Is What We Die For": Human Rights Abuses in the Democratic Republic of the Congo Power the Global Trade in Cobalt* (London: Amnesty International, 2016).

2 Andrea Murphy, Eliza Haverstock, Antoine Gara, Chris Helman, and Nathan Vardi, "Global 2000: How the World's Biggest Public Companies Endured the Pandemic," *Forbes* (May 13, 2021), https://www-statista-com.lib-e2.lib.ttu.edu/statistics/263264/top-companies-in-the-world-by-market-capitalizatison/. See Leslie Sklair, "Sleepwalking through the Anthropocene," *British Journal of Sociology* 68(4) (2017): 775–784.

3 For example, see Bill McKibben, *Eaarth: Making a Life on a Tough New Planet* (New York: Times Books, 2010); Edward O. Wilson, *Biophilia* (Cambridge: Harvard University Press, 1984); Elizabeth Kolbert, *The Sixth Extinction: An Unnatural History* (New York: Holt, 2014).

4 See Raj Patel and Jason W. Moore, *A History of the World in Seven Cheap Things: A Guide to Capitalism, Nature, and the Future of the Planet* (Oakland: University of California Press, 2017); Andreas Malm, *Fossil Capital: The Rise of Steam-Power and the Roots of Global Warming* (London: Verso, 2016); and Naomi Klein, *This Changes Everything: Capitalism vs. the Climate* (New York: Simon & Schuster, 2014).

5 Yinon M. Bar-On, Rob Phillips, and Ron Milo, "The Biomass Distribution on Earth," *Proceedings of the National Academy of Sciences* 115(25) (June 2018): 6508.

6 Vaclav Smil, "Harvesting the Biosphere: The Human Impact," *Population and Development Review* 37 (December 2011): 618.

7 Patel and Moore's clever *A History of the World in Seven Cheap Things* overlooks

the vital roles of communication and transportation in the history of capitalism.

8 Fernand Braudel proposed the idea of layers in *Civilization and Capitalism, 15th–18th Century*, vol. 1, *The Structures of Everyday Life: The Limits of the Possible*, trans. by Siân Reynolds (New York: Harper & Row, 1982), 23–26.

9 Why industrial capitalism developed in the West and not elsewhere is a question famously raised by Kenneth Pomeranz in *The Great Divergence: China, Europe, and the Making of the Modern World Economy* (Princeton: Princeton University Press, 2000). Perhaps the most important factors were environmental. In England and nowhere else, waterpower sites, coal and iron deposits, water transportation, and harbors lay very near to each other and to suitable workers.

10 Damian Carrington, "'Extraordinary' levels of pollutants found in 10km deep Mariana trench," *Guardian* (London), February 13, 2017.

1 How It Started

1 Stanley Kubrick and Arthur C. Clarke, *2001: A Space Odyssey*, film (MGM, 1988); Arthur C. Clarke, *2001: A Space Odyssey* (New York: New American Library, 1968); Michael Benson, *Space Odyssey: Stanley Kubrick, Arthur C. Clarke, and the Making of a Masterpiece* (New York: Simon & Schuster, 2018). The opening scene is dated differently in the movie and the book.

2 Sabine Gaudzinski-Windheuser, et al., "Evidence for Close-Range Hunting by Last Interglacial Neanderthals," *Nature Ecology and Evolution* 2 (July 2018): 1087–1092; Kwang Hyun Ko, "Origins of Human Intelligence: The Chain of Tool-Making and Brain Evolution," *Anthropological Notebooks* 22(1) (2016): 5–22; Vaclav Smil, *Harvesting the Biosphere: What We Have Taken from Nature* (Cambridge, Mass.: MIT Press, 2013), 74.

3 Ian Gilligan, "The Prehistoric Development of Clothing: Archaeological Implications of a Thermal Model," *Journal of Archaeological Method and Theory* 17(1) (2010): 38–39.

4 Ibid., 15–80; Ralf Kittler, Manfred Kayser, and Mark Stoneking, "Molecular Evolution of *Pediculus humanus* and the Origin of Clothing," *Current Biology* 13 (August 19, 2003): 1414–1417.

5 Chris Clarkson, et al., "Human Occupation of Northern Australia by 65,000 Years Ago," *Nature* 547 (7663) (2017): 306–310; Curtis W. Marean, "Archaeology: Early Signs of Human Presence in Australia," *Nature* 547 (7663) (2017): 285–287.

6 Eliso Kvavadze, et al., "30,000-Year-Old Wild Flax Fibers," *Science* 325 (5946) (September 11, 2009): 1359.

7 Xiaohong Wu, et al., "Early Pottery at 20,000 Years Ago in Xianrendong Cave, China," *Science* 336(6089) (June 29, 2012): 1696–1700.

8 See Jacques Ellul, *The Technological Society* (New York: Vintage Books, 1964), 24–27.
9 Rosalia Gallotti, et al., "First High Resolution Chronostratigraphy for the Early North African Acheulean at Casablanca (Morocco)." *Scientific Reports* 11(1) (2021), https://www.proquest.com/scholarly-journals/first-high-resolution-chronostratigraphy-early/docview/2555779192/se-2.
10 Robert G. Bednarik, "Early Subterranean Chert Mining," *The Artefact* 15 (1992): 11–24. Observers in 1834 saw native Tasmanians mine for ocher using hammer stones and pointed sticks; the miners were women (15).
11 Alison S. Brooks, John E. Yellen, Richard Potts, et al., "Long-Distance Stone Transport and Pigment Use in the Earliest Middle Stone Age," *Science* 360 (April 2018): 90–94. The inference is based on knowledge that in agricultural societies, and by implication long before, trade formed part of complex intergroup interactions which involved status, power, and politics.
12 Jack Goody, *Metals, Culture and Capitalism: An Essay on the Origins of the Modern World* (Cambridge: Cambridge University Press, 2012), 5; Tammy Hodgskiss, "Identifying Grinding, Scoring and Rubbing Use-Wear on Experimental Ochre Pieces," *Journal of Archaeological Science* 37(12) (December 2010): 3344–3358. See also Susan C. Vehik, "Conflict, Trade, and Political Development on the Southern Plains," *American Antiquity* 67(1) (2002): 37–64.
13 Jessica E. Tierney, Peter B. deMenocal, and Paul D. Zander, "A Climatic Context for the Out-of-Africa Migration," *Geology* 45(11) (2017): 1023–1026. The authors note that the supposed Toba bottleneck around 75,000 years ago remains in dispute (p. 1023; cp. Smil, *Harvesting the Biosphere*, 67).
14 For a detailed history of the influence of climate change on prehistoric peoples, see William James Burroughs, *Climate Change in Prehistory: The End of the Reign of Chaos* (Cambridge: Cambridge University Press, 2005).
15 Smil, *Harvesting the Biosphere*, 74.
16 Lars Werdelin and Margaret E. Lewis, "Temporal Change in Functional Richness and Evenness in the Eastern African Plio-Pleistocene Carnivoran Guild," *PLoS ONE* 8(3) (2013): e57944.
17 Felisa A. Smith, Rosemary E. Elliott Smith, S. Kathleen Lyons, and Jonathan L. Payne, "Body Size Downgrading of Mammals Over the Late Quaternary," *Science* 20 (April 2018): 311.
18 Smith, et al., "Body Size Downgrading," 310.
19 Patrick Roberts, Chris Hunt, Manuel Arroyo-Kalin, Damian Evans, and Nicole Boivin, "The Deep Human Prehistory of Global Tropical Forests and Its Relevance for Modern Conservation," *Nature Plants* 3 (2017): 17093.
20 See Richard B. Lee, "Hunter-Gatherers and Human Evolution: New Light on

Old Debates," *Annual Review of Anthropology* 47(1) (2018): 513–531, on reasons for increasing incidence of deadly conflict over time.

21 See Raymond C. Kelly, "The Evolution of Lethal Intergroup Violence," *Proceedings of the National Academy of Sciences of the United States of America* 102(43) (2005): 15294–15298; Richard W. Wrangham, "Two Types of Aggression in Human Evolution," *Proceedings of the National Academy of Sciences* 115(2) (January 2018): 245–253; and Leland Donald, "Slavery in Indigenous North America," and Neil L. Whitehead, "Indigenous Slavery in South America, 1492–1820," in David Eltis and Stanley L. Engerman, *The Cambridge World History of Slavery*, vol. 3, *AD 1420–AD 1804* (Cambridge: Cambridge University Press, 2011), 217–271.

22 Graeme Barker, *The Agricultural Revolution in Prehistory: Why Did Foragers Become Farmers?* (Oxford: Oxford University Press, 2006), 109–128. Barker thoroughly examines and judiciously evaluates all available evidence. Amaia Arranz-Otaegui, Lara Gonzalez Carretero, Monica N. Ramsey, Dorian Q. Fuller, and Tobias Richter, "Archaeobotanical Evidence Reveals the Origins of Bread 14,400 Years Ago in Northeastern Jordan," *Proceedings of the National Academy of Sciences* 115(31) (2018): 7925–7930.

23 In Central Europe, for example, Mesolithic hunter-gatherers apparently contentedly lived alongside Neolithic farmers for 2000 years. Ruth Bollongino, et al., "2000 Years of Parallel Societies in Stone Age Central Europe," *Science* 342(6157) (2013): 479–481.

24 Alfred W. Crosby, *Ecological Imperialism: The Biological Expansion of Europe* (New York: Cambridge University Press, 1986), 177.

25 Ruth Bollongino, et al., "Modern Taurine Cattle Descended from Small Number of Near-Eastern Founders," *Molecular Biology and Evolution* 29(9) (September 1, 2012): 2101–2104; Jared E. Decker, et al., "Worldwide Patterns of Ancestry, Divergence, and Admixture in Domesticated Cattle," *PLoS Genetics* 10(3) (2014): e1004254.

26 Barker, *The Agricultural Revolution in Prehistory*, 384–386.

27 R. J. Fuller, and Lu Aye, "Human and Animal Power—The Forgotten Renewables," *Renewable Energy* 48 (2012): 326–332.

28 Maria Ivanova, "The 'Green Revolution' in Prehistory: Late Neolithic Agricultural Innovations as a Technological System," in *Appropriating Innovations: Entangled Knowledge in Eurasia, 5000–1500 BCE*, ed. by Joseph Maran and Philipp Stockhammer (Oxford: Oxbow Books, 2017), 40–49.

29 Jared Diamond, *Guns, Germs, and Steel: The Fates of Human Societies* (New York: Norton, 1999), 176–191. See also Crosby, *Ecological Imperialism*, 18.

30 Melinda A. Zeder, "Domestication and Early Agriculture in the Mediterranean

Basin: Origins, Diffusion, and Impact," *Proceedings of the National Academy of Sciences* 105(33) (August 19, 2008): 11597–11604. See Crosby, *Ecological Imperialism*.

31 Lucas Stephens, Dorian Fuller, Nicole Boivin, et al., "Archaeological Assessment Reveals Earth's Early Transformation Through Land Use," *Science* 365(6456) (August 30, 2019): 897–902.

32 William F. Ruddiman, "The Anthropogenic Greenhouse Era Began Thousands of Years Ago," *Climatic Change* 61(3) (2003): 261–293. Ruddiman's thesis set off a huge debate, which he assesses in William Ruddiman, "Geographic Evidence of the Early Anthropogenic Hypothesis," *Anthropocene* 20 (2017): 4–14. He concludes that evidence has grown stronger for an anthropogenic source of Holocene warming.

33 Barker, *The Agricultural Revolution in Prehistory*, 167.

34 Svend Hansen, "Key Techniques in the Production of Metals in the 6th and 5th Millennia BCE: Prerequisites, Preconditions and Consequences," in *Appropriating Innovations: Entangled Knowledge in Eurasia, 5000–1500 BCE*, ed. by Joseph Maran and Philipp Stockhammer (Oxford: Oxbow Books, 2017), 136–148; Verena Leusch, Barbara Armbruster, Ernst Pernicka, and Vladimir Slavčev, "On the Invention of Gold Metallurgy: The Gold Objects from the Varna I Cemetery (Bulgaria)—Technological Consequence and Inventive Creativity," *Cambridge Archaeological Journal* 25(1) (February 2015): 353–376.

35 María Eugenia Aubet, *Commerce and Colonization in the Ancient Near East* (Cambridge: Cambridge University Press, 2013), 161–162.

36 Aubet, *Commerce and Colonization*, 157–199, 222–223, 236–238, 283–296, 336–343.

37 On ancient Mesopotamian capitalism, see Michael Jursa, "Babylonia in the First Millennium BCE—Economic Growth in Times of Empire," in *The Cambridge History of Capitalism*, ed. by Larry Neal and Jeffrey G. Williamson, vol. 1 (Cambridge: Cambridge University Press, 2014), 24–42.

38 George Modelski, *World Cities: 3000 to 2000* (Washington, D.C.: Faros 2000, 2003), 20–32.

39 Jianjun Mei, Yongbin Yu, Kunlong Chen, and Lu Wang, "The Appropriation of Early Bronze Technology in China," in *Appropriating Innovations: Entangled Knowledge in Eurasia, 5000–1500 BCE*, ed. by Joseph Maran, and Philipp Stockhammer (Oxford: Oxbow Books, 2017).

40 Elizabeth C. Stone, "The Trajectory of Social Inequality in Ancient Mesopotamia," and Timothy A. Kohler, et al., "Deep Inequality: Summary and Conclusions," in *Ten Thousand Years of Inequality: The Archaeology of Wealth*

Differences, ed. by Timothy A. Kohler and Michael Ernest Smith (Tucson: University of Arizona Press, 2018), 230–261, 289–218.

41 Richard W. Yerkes and Ran Barkai, "Tree-Felling, Woodworking, and Changing Perceptions of the Landscape during the Neolithic and Chalcolithic Periods in the Southern Levant," *Current Anthropology* 54(2) (April 2013): 222–231.

42 J. P. Grattan, D. D. Gilbertson, and C. O. Hunt, "The Local and Global Dimensions of Metalliferous Pollution Derived from a Reconstruction of an Eight-Thousand-Year Record of Copper Smelting and Mining at a Desert-Mountain Frontier in Southern Jordan," *Journal of Archaeological Science* 34 (2007): 83–110.

43 Sing C. Chew, "Ecological Relations and the Decline of Civilizations in the Bronze Age World-System: Mesopotamia and Harappa 2500 BC–1700 BC," in *Ecology and the World-System*, ed. by Walter L. Goldfrank, David Goodman, and Andrew Szasz (Westport, Conn.: Greenwood Press, 1999), 87–106.

44 In 2018, the International Commission on Stratigraphy officially ratified this event as the beginning of the Meghalayan, the last age of the Holocene.

45 David M. Schaps, *The Invention of Coinage and the Monetization of Ancient Greece* (Ann Arbor: University of Michigan Press, 2010).

46 Alain Bresson, "Capitalism and the Ancient Greek Economy," in *The Cambridge History of Capitalism*, ed. by Larry Neal and Jeffrey G. Williamson, vol. 1 (Cambridge: Cambridge University Press, 2014), 43–74. See Edmund S. Morgan, *American Slavery, American Freedom: The Ordeal of Colonial Virginia* (New York: Norton, 1975).

47 Willem M. Jongman, "Re-Constructing the Roman Economy," in *The Cambridge History of Capitalism*, ed. by Larry Neal and Jeffrey G. Williamson, vol. 1 (Cambridge: Cambridge University Press, 2014), 75–100.

48 A. H. V. Smith, "Provenance of Coals from Roman Sites in England and Wales," *Britannia* 28 (1997): 297–324; Martin J. Dearne and Keith Branigan. "The Use of Coal in Roman Britain," *Antiquaries Journal* 75 (1995): 71–105.

49 Modelski, *World Cities*, 39–59; Walter Scheidel, "Demography," in *The Cambridge Economic History of the Greco-Roman World*, ed. by Walter Scheidel, Ian Morris, and Richard P. Saller (Cambridge: Cambridge University Press, 2007), 38–86.

50 J. Donald Hughes, *Pan's Travail: Environmental Problems of the Ancient Greeks and Romans* (Baltimore: Johns Hopkins University Press, 1994), 149–168; Dearne and Branigan, "The Use of Coal in Roman Britain," 86–87.

51 Kyle Harper, *The Fate of Rome: Climate, Disease, and the End of an Empire* (Princeton: Princeton University Press, 2017).

52 Frederic L. Cheyette, "The Disappearance of the Ancient Landscape and the

Climatic Anomaly of the Early Middle Ages: A Question to Be Pursued," *Early Medieval Europe* 16(2) (2008): 127–165.

53 Arie S. Issar and Mattanyah Zohar, *Climate Change: Environment and Civilization in the Middle East* (Berlin: Springer, 2004), 165–176, 212–219, 226–228.

54 R. B. Wong, "China before Capitalism," 125–164; and Karl Gunnar Persson, "Markets and Coercion in Medieval Europe," 225–235; both in *The Cambridge History of Capitalism*, ed. by Larry Neal and Jeffrey G. Williamson, vol. 1 (Cambridge: Cambridge University Press, 2014).

2 *Trade and Empire*

1 Bartolomé de las Casas, quoted in Rebecca Catz, *Christopher Columbus and the Portuguese, 1476–1498* (Westport, Conn.: Greenwood Press, 1993), 12.

2 Henry Kamen, *Spain, 1469–1714: A Society of Conflict*, 3rd edn. (Harlow, England: Pearson/Longman, 2005), 16.

3 William D. Phillips and Carla Rahn Phillips, *The Worlds of Christopher Columbus* (Cambridge: Cambridge University Press, 1992), 134.

4 Sven Beckert, *Empire of Cotton: A Global History* (New York: Knopf, 2014), 9.

5 Columbus, quoted in Phillips and Phillips, *Worlds of Christopher Columbus*, 183–185.

6 See Alfred W. Crosby, *The Columbian Exchange: Biological and Cultural Consequences of 1492* (Westport, Conn.: Greenwood Press, 1972); Crosby, *Ecological Imperialism*.

7 Phillips and Phillips, *Worlds of Christopher Columbus*, 194–211, 223–224, 96.

8 Brian Graham, "The Mediterranean in the Medieval and Renaissance World," in *The Mediterranean: Environment and Society*, ed. by Russell King, L. J Proudfoot, and Bernard J. Smith (London: Arnold, 1997); on Italian capitalism, especially Venetian, see Fernand Braudel, *Civilization and Capitalism, 15th–18th Century*, vol. 3, *Perspective of the World*, trans. by Siân Reynolds (New York: Harper & Row, 1982), 116–138.

9 Gino Luzzatto, *Breve Storia Economica dell'Italia Medievale: Dalla Caduta dell'Impero Romano al Principio del Cinquecento* (Torino: Einaudi, 1966), 137–141.

10 Patrick McCray, *Glassmaking in Renaissance Venice: The Fragile Craft* (Aldershot, Hants, England: Ashgate, 1999), 136; Steven Epstein, *Genoa and the Genoese, 958–1528* (Chapel Hill: University of North Carolina Press, 1996), 275; Helen Nader, "Desperate Men, Questionable Acts: The Moral Dilemma of Italian Merchants in the Spanish Slave Trade," *Sixteenth Century Journal* 33(2) (2002): 405–406.

11 Adam Smith, *The Wealth of Nations*, Book 1, Chapter 1.
12 François Menant, *L'Italie des Communes: 1100–1350* (Paris: Belin, 2005), 283–286.
13 Braudel, *Civilization and Capitalism*, vol. 3, 550–552; Carlo Poni, "The Circular Silk Mill: A Factory Before the Industrial Revolution in Early Modern Europe," *History of Technology* 21 (1999): 65–85. Technical challenges for silk thread were not as difficult as those for making cotton thread, though.
14 William Gervase Clarence-Smith and David Eltis, "White Servitude," in *The Cambridge World History of Slavery*, vol. 3, *AD 1420–AD 1804*, ed. by David Eltis and Stanley L. Engerman (Cambridge: Cambridge University Press, 2011), 132.
15 Daniel Hershenzon, *The Captive Sea: Slavery, Communication, and Commerce in Early Modern Spain and the Mediterranean* (Philadelphia: University of Pennsylvania Press, 2018), 2.
16 Epstein, *Genoa and the Genoese*, 266–270; and Michel Balard, "Slavery in the Latin Mediterranean (Thirteenth to Fifteenth Centuries): The Case of Genoa," and Danuta Quirini-Poplawska, "The Venetian Involvement in the Black Sea Slave Trade (Fourteenth to Fifteenth Centuries)," in *Slavery and the Slave Trade in the Eastern Mediterranean (c.1000–1500 CE)*, ed. by Reuven Amitai and Christoph Cluse (Turnhout, Belgium: Brepols, 2017).
17 Thomas Allison Kirk, *Genoa and the Sea: Policy and Power in an Early Modern Maritime Republic, 1559–1684* (Baltimore: Johns Hopkins University Press, 2005), 12.
18 W. Rothwell, "Sugar and Spice and All Things Nice: From Oriental Bazar to English Cloister in Anglo-French," *Modern Language Review* 100 (2005): 38–50; Stefan Halikowski Smith, "Demystifying a Change in Taste: Spices, Space, and Social Hierarchy in Europe, 1380–1750," *International History Review* 29(2) (June 2007): 237–257.
19 J. H. Galloway, "The Mediterranean Sugar Industry," *Geographical Review* 67(2) (April 1977): 179–190.
20 Ibid., 188–189.
21 William D. Phillips Jr., "Sugar in Iberia," in *Tropical Babylons: Sugar and the Making of the Atlantic World, 1450–1680*, ed. by Stuart B. Schwartz (Chapel Hill: University of North Carolina Press, 2004), 28–31; Adela Fabregas-Garcia, "Commercial Crop or Plantation System? Sugar Cane Production from the Mediterranean to the Atlantic," in *From Al-Andalus to the Americas (13th–17th Centuries): Destruction and Construction of Societies*, ed. by Thomas F. Glick, Antonio Malpica Cuello, Fèlix Retamero, and Josep Torró Abad (Leiden: Brill, 2018), 301–307.
22 Nicholas Coureas, "Hospitaller Estates and Agricultural Production on Fourteenth- and Fifteenth-Century Cyprus," in *Islands and Military Orders,*

c.1291–c.1798, ed. by Emanuel Buttigieg and Simon Phillips (London: Taylor and Francis, 2016), 215–224.

23 Arie S. Issar and Mattanyah Zohar, *Climate Change: Environment and Civilization in the Middle East* (Berlin: Springer, 2004), 165–176, 212–219, 226–228.

24 Marina Solomidou-Ieronymidou, "Sugar Mills and Sugar Production in Medieval Cyprus," *Medieval Cyprus: A Place of Cultural Encounter: Conference in Münster, 6–8 December 2012*, ed. by Sabine Rogge and Michael Grünbart (Münster: Waxmann, 2015), 147–173.

25 Juan Garcia Latorre, Andrés Sánchez Picón, and Jesús García Latorre, "The Man-Made Desert: Effects of Economic and Demographic Growth on the Ecosystems of Arid Southeastern Spain," *Environmental History* 6(1) (January 2001): 75–94.

26 Gonzalo Anes, "The Agrarian 'Depression' in Castile in the Seventeenth Century," and Enrique Llopis Agelán, "Castilian Agriculture in the Seventeenth Century: Depression, or 'Readjustment and Adaptation'?," in *The Castilian Crisis of the Seventeenth Century: New Perspectives on the Economic and Social History of Seventeenth-Century Spain*, ed. by I. A. A. Thompson and Bartolomé Yun Casalilla (Cambridge: Cambridge University Press, 1994), 60–76, 77–100; María Valbuena-Carabaña, Unai López de Heredia, Pablo Fuentes-Utrilla, Inés González-Doncel, and Luis Gil, "Historical and Recent Changes in the Spanish Forests: A Socio-Economic Process," *Review of Palaeobotany and Palynology* 162(3) (October 2010): 496–497; Richard C. Hoffmann, *An Environmental History of Medieval Europe* (Cambridge: Cambridge University Press, 2014), 176–180.

27 Karl W. Butzer and Sarah E. Harris, "Geoarchaeological Approaches to the Environmental History of Cyprus: Explication and Critical Evaluation," *Journal of Archaeological Science* 34(11) (November 2007): 1932–1952.

28 McCray, *Glassmaking in Renaissance Venice*, 145–146.

29 Ibid., 101–114; David Jacoby, "Raw Materials for the Glass Industries of Venice and the Terraferma, about 1370–about 1460," *Journal of Glass Studies* 35 (1993): 65–90.

30 Harald Thomasius, "The Influence of Mining on Woods and Forestry in the Saxon Erzgebirge up to the Beginning of the 19th Century," *GeoJournal* 32(2) (1994): 122–123. See also Saúl Guerrero, *Silver by Fire, Silver by Mercury: A Chemical History of Silver Refining in New Spain and Mexico, 16th to 19th Centuries* (Leiden: Brill, 2017), 78–101.

31 Thomasius, "Influence of Mining," 113–119; Boško Bojović, "Entre Venise et L'Empire Ottoman: Les Métaux Précieux des Balkans (XVe–XVIe Siècle)," *Annales.*

Histoire, Sciences Sociales 60(6) (2005): 1277–1297; Elisabeth Breitenlechner, Marina Hilber, Joachim Lutz, Yvonne Kathrein, Alois Unterkircher, and Klaus Oeggl, "Reconstructing the History of Copper and Silver Mining in Schwaz, Tirol," *RCC Perspectives* 10 (2012): 7–20; Laura Hollsten, "Mercurial Activity and Subterranean Landscapes: Towards an Environmental History of Mercury Mining in Early Modern Idrija," *RCC Perspectives* 10 (2012): 21–38; Jeannette Graulau, "Finance, Industry and Globalisation in the Early Modern Period: The Example of the Metallic Business of the House of Fugger," *Rivista di Studi Politici Internazionali* 75(4) (300) (2008): 554–598.

32 Francis N. N. Botchway, "Pre-Colonial Methods of Gold Mining and Environmental Protection in Ghana," *Journal of Energy and Natural Resources Law* 13(4) (1995): 299–311; Eugenia W. Herbert, "Elusive Frontiers: Precolonial Mining in Sub-Saharan Africa," in *Mining Frontiers in Africa: Anthropological and Historical Perspectives*, ed. by Werthmann, Katja, and Tilo Grätz (Cologne, Germany: Rüdiger Köppe, 2012), 23–32.

33 A. R. Disney, *Twilight of the Pepper Empire: Portuguese Trade in Southwest India in the Early Seventeenth Century* (Cambridge: Harvard University Press, 1978), 32–36; John Keay, *The Spice Route: A History* (Berkeley: University of California Press, 2006), 215.

34 Ruth Pike, *Enterprise and Adventure: The Genoese in Seville and the Opening of the New World* (Ithaca, N.Y.: Cornell University Press, 1966), 48.

35 Count of Osorono to the Crown (1523), quoted in Pike, *Enterprise and Adventure*, 19.

36 Possibly Corsica: Karl Appuhn, personal communication, September 14, 2018.

37 Phillips, "Sugar in Iberia," 31–34.

38 William D. Phillips, *Slavery in Medieval and Early Modern Iberia* (Philadelphia: University of Pennsylvania Press, 2014), 10–27; Epstein, *Genoa and the Genoese*, 310–311.

39 Catz, *Christopher Columbus*, 1–9; David B. Quinn, "Columbus and the North: England, Iceland, and Ireland," *William and Mary Quarterly* 49(2) (1992): 278–297.

40 Ibid., 293.

41 Pike, *Enterprise and Adventure*, 48–83, 99–100; Delno West, "Christopher Columbus and his Enterprise to the Indies: Scholarship of the Last Quarter Century," *William and Mary Quarterly* 49(2) (1992): 260–261; Phillips and Phillips, *Worlds of Christopher Columbus*, 230. See also D. B. Quinn, "The Italian Renaissance and Columbus," *Renaissance Studies* 6(3/4) (1992): 352–359.

42 Nader, "Desperate Men, Questionable Acts," 401–404, 407–409, 415–421.

43 Antonio de Almeida Mendes, "Portugal, Morocco and Guinea: Reconfiguration

of the North Atlantic at the End of the Middle Ages," in *From Al-Andalus to the Americas (13th–17th Centuries): Destruction and Construction of Societies*, ed. by Thomas F. Glick, Antonio Malpica, Fèlix Retamero, and Josep Torró (Leiden: Brill, 2018), 405.

44 For a different perspective on Portuguese expansion, see Braudel, *Civilization and Capitalism*, vol. 3, 138–143.

45 See J. H. Parry, *The Age of Reconnaissance* (Cleveland, Ohio: World, 1963), 53–68, 83–113.

46 De Almeida Mendes, "Portugal, Morocco and Guinea," 407–408.

47 Anna Unali, *Alla Ricerca dell'Oro: Mercanti, Viaggiatori, Missionari in Africa e nelle Americhe (secc. XIII–XVI)* (Roma: Bulzoni, 2006), 207–250; Braudel, *Civilization and Capitalism*, vol. 3, 430–441.

48 Alberto Vieira, "Sugar Islands: The Sugar Economy of Madeira and the Canaries, 1450–1650," in Stuart B. Schwartz, *Tropical Babylons: Sugar and the Making of the Atlantic World, 1450–1680* (Chapel Hill: University of North Carolina Press, 2004), 42–43.

49 Kirk, *Genoa and the Sea*, 15.

50 António de Almeida Mendes, "Les Réseaux de la Traite Ibérique dans l'Atlantique Nord (1440–1640)," *Annales. Histoire, Sciences Sociales* 63(4) (2008): 755.

51 Vieira, "Sugar Islands," in Schwartz, *Tropical Babylons*, 42–84; J. H. Galloway, *The Sugar Cane Industry: An Historical Geography from Its Origins to 1914* (Cambridge: Cambridge University Press, 1989), 50–55; Alberto Vieira, *Os Escravos no Arquipélago da Madeira: Séculos XV a XVII* (Funchal, Madeira: Centro de Estudos de História do Atlântico, 1991).

52 Catz, *Christopher Columbus*, 21–36; Vieira, *Os Escravos.*

53 Genero Rodríguez Morel, "The Sugar Economy of Española in the Sixteenth Century," in Schwartz, *Tropical Babylons*, 87, 92–93; Galloway, *Sugar Cane Industry*, 55–58.

54 Arlindo Manuel Caldeira, "Learning the Ropes in the Tropics: Slavery and the Plantation System on the Island of São Tomé," *African Economic History* 39 (2011): 36–37.

55 Caldeira, "Learning the Ropes," 37–43, 45.

56 De Almeida Mendes, "Les Réseaux," 756; Caldeira, "Learning the Ropes," 50–51.

57 Paul E. Lovejoy, *Transformations in Slavery: A History of Slavery in Africa*, 3rd edn. (Cambridge: Cambridge University Press, 2011), 42.

58 H. A. Gemery and J. S. Hogendorn, "Comparative Disadvantage: The Case of Sugar Cultivation in West Africa," *Journal of Interdisciplinary History* 9(3) (1979): 429–449.

59 Caldeira, "Learning the Ropes," 51–63; Galloway, *Sugar Cane Industry*, 58–61.

60 Barbara L. Solow, "Slavery and Colonization," in *Slavery and the Rise of the Atlantic System*, ed. by Barbara Solow (Cambridge: Cambridge University Press, 1991). Jason Moore ascribes a much more prominent place to Madeira, which I present as a way-station between the Italian Mediterranean model and the full-blown sugar plantations of São Tomé. See Jason W. Moore, "Madeira, Sugar, and the Conquest of Nature in the 'First' Sixteenth Century: Part I: From 'Island of Timber' to Sugar Revolution, 1420–1506," *Review (Fernand Braudel Center)* 32(4) (2009): 345–390; and "Madeira, Sugar, and the Conquest of Nature in the 'First' Sixteenth Century, Part II: From Regional Crisis to Commodity Frontier, 1506–1530," *Review (Fernand Braudel Center)* 33(1) (2010): 1–24. His argument rests on the pace of deforestation.

61 Cortez, quoted in David Watts, *The West Indies: Patterns of Development, Culture, and Environmental Change Since 1492* (Cambridge: Cambridge University Press, 1987), 78.

62 Ibid., 71–72.

63 Antonio de Almeida Mendes, "Portugal, Morocco and Guinea," 412.

64 Watts, *West Indies*, 121–127; Galloway, *Sugar Cane Industry*, 61–70.

65 Gilberto Freyre, *The Masters and the Slaves (Casa-Grande and Senzala): A Study in the Development of Brazilian Civilization*, 2nd English language edn., rev edn. (New York: Knopf, 1956), 12–77; Arnold Wiznitzer, "The Jews in the Sugar Industry of Colonial Brazil," *Jewish Social Studies* 18(3) (1956): 189–198.

66 See Barbara L. Solow, "Slavery and Colonization."

67 João Fragoso and Ana Rios, "Slavery and Politics in Colonial Portuguese America: The Sixteenth to the Eighteenth Centuries," in Eltis and Engerman, *Cambridge World History of Slavery*, vol. 3, 350; Galloway, *Sugar Cane Industry*, 70–77, 83; de Almeida Mendes, "Les Réseaux," 764–766. See also Andrés Reséndez, *The Other Slavery: The Uncovered Story of Indian Enslavement in America* (Boston: Houghton Mifflin Harcourt, 2016).

68 Warren Dean, *With Broadax and Firebrand: The Destruction of the Brazilian Atlantic Forest* (Berkeley: University of California Press, 1995), 1–90; Thomas D. Rogers, *The Deepest Wounds: A Labor and Environmental History of Sugar in Northeast Brazil* (Chapel Hill: University of North Carolina Press, 2010), 21–36; Shawn William Miller, *Fruitless Trees: Portuguese Conservation and Brazil's Colonial Timber* (Stanford: Stanford University Press, 2000), 9.

69 Alonso de la Mota y Escobar, quoted in P. J. Bakewell, *Silver Mining and Society in Colonial Mexico: Zacatecas, 1546–1700* (Cambridge: Cambridge University Press, 1971), 1.

70 Guerrero, *Silver by Fire*, 9–12.

71 Nicholas A. Robins, *Mercury, Mining, and Empire: The Human and Ecological Cost of Colonial Silver Mining in the Andes* (Bloomington: Indiana University Press, 2011).

72 See also Kris E. Lane, *Potosí: The Silver City That Changed the World* (Oakland: University of California Press, 2019).

73 Tetsuya Ogura et al., "Zacatecas (Mexico) Companies Extract Hg from Surface Soil Contaminated by Ancient Mining Industries," *Water, Air, and Soil Pollution* 148 (2003): 167–177; Irma Gavilán-García et. al., "Mercury Speciation in Contaminated Soils from Old Mining Activities in Mexico Using a Chemical Selective Extraction," *Journal of the Mexican Chemical Society* 52(4) (October–December 2008): 263–271; D. A. De la Rosa et al., "Survey of Atmospheric Total Gaseous Mercury in Mexico," *Atmospheric Environment* 38(29) (September 2004): 4839–4846.

74 See Bakewell, *Silver Mining and Society in Colonial Mexico*, 1–173; Guerrero, *Silver by Fire*, 75–101, 316–360.

75 Bakewell, *Silver Mining and Society in Colonial Mexico*, 173–186; Robins, *Mercury, Mining, and Empire*, 29–31; Stanley J. Stein and Barbara H. Stein, *Silver, Trade, and War: Spain and America In the Making of Early Modern Europe* (Baltimore: Johns Hopkins University Press, 2000), 260–266.

76 Elinor G. K. Melville, *A Plague of Sheep: Environmental Consequences of the Conquest of Mexico* (Cambridge: Cambridge University Press, 1994); Richard J. Salvucci, *Textiles and Capitalism in Mexico: An Economic History of the Obrajes, 1539–1840* (Princeton: Princeton University Press, 1987).

77 Stein and Stein, *Silver, Trade, and War*, 266.

78 Braudel, *Civilization and Capitalism*, vol. 3, 157–174.

79 Dean, *Broadax and Firebrand*, 91–99.

80 Antonio de Almeida Mendes, "Portugal, Morocco and Guinea," 402–403.

81 Harry E. Cross, "South American Bullion Production and Export, 1550–1750," in *Precious Metals in the Later Medieval and Early Modern Worlds*, ed. by John F. Richards (Durham, N.C.: Carolina Academic Press, 1983), 418, 420.

3 The Wonders of Coal and Machines

1 A. R. Mitchell, "The European Fisheries in Early Modern History," in *The Cambridge Economic History of Europe*, vol. 5, *The Economic Organisation of Early Modern Europe*, ed. by E. E. Rich and Charles Wilson (Cambridge: Cambridge University Press, 1977), 147–148.

2 Paul Crutzen and Eugene F. Stoermer, "Opinion: Have we entered the 'Anthropocene'?" http://www.igbp.net/news/opinion/opinion/haveweenteredtheanthropocene.5.d8b4c3c12bf3be638a8000578.html (October 31, 2010).

3 Tom Boden, Bob Andres, and Gregg Marland, "Global CO2 Emissions from Fossil-Fuel Burning, Cement Manufacture, and Gas Flaring: 1751–2014," Carbon Dioxide Information Analysis Center, Oak Ridge National Laboratory, Oak Ridge, Tennessee, March 3, 2017, https://cdiac.ess-dive.lbl.gov/ftp/ndp030/global.1751_2014.ems.

4 E. Patricia Dennison, "Glasgow to 1700," in *The Oxford Companion to Scottish History*, ed. by Michael Lynch (Oxford: Oxford University Press, 2001), 267.

5 Gabri van Tussenbroek, "The Great Rebuilding of Amsterdam (1521–1578)," *Urban History* 46(3) (August 2019): 427; Liên Luu, *Immigrants and the Industries of London, 1500–1700* (Abingdon, Oxon: Routledge, 2016), 34.

6 Jan Bieleman, *Five Centuries of Farming: A Short History of Dutch Agriculture, 1500–2000* (Wageningen, Netherlands: Wageningen Academic Publishers, 2010), 35–76; Guus J. Borger and Willem A. Ligtendag, "The Role of Water in the Development of the Netherlands: A Historical Perspective," *Journal of Coastal Conservation* 4(2) (1998): 109–114; William H. TeBrake, "Taming the Waterwolf: Hydraulic Engineering and Water Management in the Netherlands during the Middle Ages," *Technology and Culture* 43(3) (2002): 475–499; Arne Kaijser, "System Building from Below: Institutional Change in Dutch Water Control Systems," *Technology and Culture* 43(3) (2002): 521–548; Petra J. E. M. Van Dam, "Ecological Challenges, Technological Innovations: The Modernization of Sluice Building in Holland, 1300–1600," *Technology and Culture* 43(3) (2002): 500–520; Jan De Vries and A. M. Van Der Woude, *The First Modern Economy: Success, Failure, and Perseverance of the Dutch Economy, 1500–1815* (Cambridge: Cambridge University Press, 1997), 195–210; Fernand Braudel, *Civilization and Capitalism, 15th–18th Century*, vol. 3, *Perspective of the World*, trans. by Siân Reynolds (New York: Harper & Row, 1982), 175–206.

7 De Vries and Van Der Woude, *First Modern Economy*, 329.

8 Dagomar Degroot, *The Frigid Golden Age: Climate Change, the Little Ice Age, and the Dutch Republic, 1560–1720* (Cambridge: Cambridge University Press, 2018), 75–79.

9 Robert Greenhalgh Albion, *Forests and Sea Power: The Timber Problem of the Royal Navy, 1652–1862* (Cambridge: Harvard University Press, 1926), 139–156.

10 Degroot, *Frigid Golden Age.*

11 Robert Siebelhoff, "The Demography of the Low Countries 1500–1990: Facts and figures," *Canadian Journal of Netherlandic Studies* 14(1) (Spring 1993): 127.

12 On Antwerp's rise and fall, see Braudel, *Civilization and Capitalism*, vol. 3, 143–157.

13 De Vries and Van Der Woude, *First Modern Economy*, 270–349, 665–710; Braudel, *Civilization and Capitalism*, vol. 3, 235–276.

14 Edmund S. Morgan, *American Slavery, American Freedom: The Ordeal of Colonial Virginia* (New York: Norton, 1975), 28–31, 85–87; Carville Earle, *Geographical Inquiry and American Historical Problems* (Stanford: Stanford University Press, 1992), 67–69.
15 DeGroot, *Frigid Golden Age*, 244–245.
16 B. H. Slicher van Bath, "Agriculture and the Vital Revolution," in *The Cambridge Economic History of Europe*, vol. 5, *The Economic Organisation of Early Modern Europe*, ed. by E. E. Rich and Charles Wilson (Cambridge: Cambridge University Press, 1977), 110–113.
17 Margrit Schulte Beerbühl et al., "War England ein Sonderfall der Industrialisierung? Der ökonomische Einfluß der Protestantischen Immigranten auf die Entwicklung der Englischen Wirtschaft vor der Industrialisierung," *Geschichte und Gesellschaft* 21(4) (1995): 479–505; Sydney John Chapman, *The Lancashire Cotton Industry: A Study in Economic Development* (Manchester: Manchester University Press, 1904), 1; Braudel, *Civilization and Capitalism*, vol. 3, 558–564.
18 John Robert McNeill, *Mosquito Empires: Ecology and War in the Greater Caribbean, 1620–1914* (New York: Cambridge University Press, 2010), 137–191.
19 Pieter C. Emmer and Stanley L. Engerman, "The Non-Hispanic West Indies," in *The Cambridge World History of Slavery*, vol. 4, *AD 1804–AD 2016*, ed. by David Eltis, Stanley L. Engerman, Seymour Drescher, and David Richardson (Cambridge: Cambridge University Press, 2017), 73. The venerable, foundational study of sugar is Sidney W. Mintz, *Sweetness and Power: The Place of Sugar in Modern History* (New York: Viking, 1985).
20 Ralph A. Austen and Woodruff D. Smith, "Private Tooth Decay as Public Economic Virtue: The Slave–Sugar Triangle, Consumerism, and European Industrialization," *Social Science History* 14(1) (1990): 98–102. For an assessment of Europe's economic relationship with the Americas, see Braudel, *Civilization and Capitalism*, vol. 3, 387–429.
21 Morgan, *American Slavery, American Freedom*, 295–315; Philip D. Morgan, "Virginia Slavery in Atlantic Context, 1550 to 1650," in *Virginia 1619: Slavery and Freedom in the Making of English America*, ed. by Paul Musselwhite, Peter C. Mancall, and James Horn (Chapel Hill: University of North Carolina Press, 2019), 85.
22 See, for example, John Hughes (later Catholic archbishop of New York) in Revd. John Hughes and Revd. John Breckinridge, *A Discussion of the Question, Is the Roman Catholic Religion, in Any or All Its Principles or Doctrines, Inimical to Civil or Religious Liberty?: And of the Question, Is the Presbyterian Religion,*

in Any or All Its Principles or Doctrines, Inimical to Civil or Religious Liberty? (Philadelphia: Carey, Lea, and Blanchard, 1836), 391.

23 Trevor G. Burnard, *Planters, Merchants, and Slaves: Plantation Societies in British America, 1650–1820* (Chicago: University of Chicago Press, 2015), 55.

24 Quoted in Iain Whyte, *Scotland and the Abolition of Black Slavery, 1756–1838* (Edinburgh: Edinburgh University Press, 2006), 61.

25 Richard K. Fleischman, David Oldroyd, and Thomas N. Tyson, "Plantation Accounting and Management Practices in the US and the British West Indies at the End of Their Slavery Eras," *Economic History Review* 64(3) (2011): 765–797; Caitlin Rosenthal, *Accounting for Slavery: Masters and Management* (Cambridge: Harvard University Press, 2018).

26 Kenneth M. Stampp, *The Peculiar Institution: Slavery in the Ante-Bellum South* (New York: Knopf, 1956), 192–236; Winthrop D. Jordan, *White over Black: American Attitudes toward the Negro, 1550–1812* (Chapel Hill: University of North Carolina Press, 1968), 3–135, 204–212; Stanley M. Elkins, *Slavery: A Problem in American Institutional and Intellectual Life*, 3rd edn. (Chicago: University of Chicago Press, 1976), 52–80; Eugene D. Genovese, *Roll, Jordan, Roll: The World the Slaves Made* (New York: Pantheon Books, 1974), 88–93; Philip J. Greven, *The Protestant Temperament: Patterns of Child-Rearing, Religious Experience, and the Self in Early America* (New York: Knopf, 1977), 32–42. The complex interaction between culture, morality, capitalism, and racial prejudice receives discussion in Robin Blackburn, *The Making of New World Slavery: From the Baroque to the Modern, 1492–1800* (London: Verso, 1997), 350–363.

27 Philip D. Morgan, "Slavery in the British Caribbean," in *The Cambridge World History of Slavery*, vol. 3, *AD 1420–AD 1804*, ed. by David Eltis and Stanley L. Engerman (Cambridge: Cambridge University Press, 2011), 378–406.

28 Seymour Drescher, "The Long Goodbye: Dutch Capitalism and Antislavery in Comparative Perspective," *American Historical Review* 99(1) (1994): 44–69. Curiously, in stark contrast with Britain and America, no significant abolitionist movement ever arose in the Netherlands.

29 Blackburn, *New World Slavery*, 188–195, 362–363.

30 Whyte, *Scotland and the Abolition of Black Slavery*, 9–40. A truly consistent Calvinist who defended slavery could not accept it uncritically. See Marilyn J. Westerkamp, "James Henley Thornwell, Pro-Slavery Spokesman within a Calvinist Faith," *The South Carolina Historical Magazine* 87(1) (1986): 49–64; and Tommy W. Rogers, "Dr. Frederick A. Ross and the Presbyterian Defense of Slavery," *Journal of Presbyterian History* 45(2) (1967): 112–124.

31 Marie Houllemare, "Procedures, Jurisdictions and Records: Building the French

Empire in the Early Eighteenth Century," *Journal of Colonialism and Colonial History* 21(2) (2020): 13.

32 Galloway, *Sugar Cane Industry*, 102, 103.

33 Ibid., 103–104; Watts, *West Indies*, 211.

34 Galloway, *Sugar Cane Industry*, 74–77.

35 David Watts, *West Indies*, 219–223.

36 Ibid., 222.

37 Ibid., 400–401.

38 Galloway, *Sugar Cane Industry*, 94–105; Watts, *West Indies*, 401–403, 411.

39 Richard Grove, *Green Imperialism: Colonial Expansion, Tropical Island Edens, and the Origins of Environmentalism, 1600–1860* (Cambridge: Cambridge University Press, 1995), esp. 153–308.

40 Quoted in Burnard, *Planters, Merchants, and Slaves*, 55–56.

41 Ibid., 20. "Anthropocene" has inspired a proliferation of names ending in "-cene," including the awkward "capitalocene," coined by Andreas Malm and promoted by Jason Moore, and the even more ungainly "plantationocene," coined in 2014 and championed (perhaps tongue in cheek) by Donna Haraway. See Donna Haraway, Noboru Ishikawa, Scott F. Gilbert, Kenneth Olwig, Anna L. Tsing, and Nils Bubandt, "Anthropologists are Talking—About the Anthropocene," *Ethnos* 81(3) (2016): 535–564. Haraway et al. do not include the steam engine in their discussion of the environmental impact of the plantation.

42 Albert E. Cowdrey, *This Land, This South: An Environmental History* (Lexington: University Press of Kentucky, 1983), 24–63; Timothy Silver, *A New Face on the Countryside: Indians, Colonists, and Slaves in the South Atlantic Forests, 1500–1800* (Cambridge: Cambridge University Press, 1990), 67–185.

43 On the Carolinas, see Peter H. Wood, *Black Majority: Negroes in Colonial South Carolina from 1670 through the Stono Rebellion* (New York: Knopf, 1974); Mart A. Stewart, *"What Nature Suffers to Groe": Life, Labor, and Landscape on the Georgia Coast, 1680–1920* (Athens: University of Georgia Press, 1996). On African origins of rice, see David Eltis, Philip Morgan, and David Richardson, "Black, Brown, or White? Color-Coding American Commercial Rice Cultivation with Slave Labor," *American Historical Review* 115(1) (2010): 164–171.

44 Roger Emerson, "The Contexts of the Scottish Enlightenment," in *The Cambridge Companion to the Scottish Enlightenment*, ed. by Alexander Broadie (Cambridge: Cambridge University Press, 2003), 9–30.

45 Christopher Smout, "The Culture of Migration: Scots as Europeans 1500–1800," *History Workshop Journal* 40 (1995): 108–117.

46 John M. Mackenzie, "Essay and Reflection: On Scotland and the Empire," *International History Review* 15(4) (1993): 721–728.

47 Alan McKinlay and Alistair Mutch, "'Accountable Creatures': Scottish Presbyterianism, Accountability and Managerial Capitalism," *Business History* 57(2) (2015): 241–256.

48 See T. M. Devine, "The Colonial Trades and Industrial Investment in Scotland, c.1700–1815," *Economic History Review* 29(1) (1976): 1–13.

49 Ibid., 1; J. E. Inikori, *Africans and the Industrial Revolution in England: A Study in International Trade and Economic Development* (New York: Cambridge University Press, 2002), 330.

50 "Alexander Macfarlane," *Legacies of British Slave-ownership database*, https://www.ucl.ac.uk/lbs/person/view/2146644157.

51 See Ben Marsden, *Watt's Perfect Engine: Steam and the Age of Invention* (New York: Columbia University Press, 2002), 9–16; James Black to Watt, and John Robison to Watt, in James Watt and Joseph Black, *Partners in Science: Letters of James Watt and Joseph Black*, ed. by Eric Robinson and Douglas McKie (Cambridge: Harvard University Press, 1970), 253–257.

52 James Patrick Muirhead, *The Life of James Watt: With Selections from His Correspondence*, 2nd rev. edn. (London: John Murray, 1859), 46–49.

53 Richard Leslie Hills, *Power from Steam: A History of the Stationary Steam Engine* (Cambridge: Cambridge University Press, 1989), 16–30.

54 Hills, *Power from Steam*, 51–54, 59.

55 Andrew Carnegie, *James Watt* (Edinburgh: Oliphant, Anderson, and Ferrier, 1903), 7–8, 13, 138–139, 143; George Williamson, *Memorials of the Lineage, Early Life, Education, and Development of the Genius of James Watt* (Edinburgh: Constable, 1856), 41n1, 46, 48–49, 94–101, 147.

56 Muirhead, *The Life of James Watt*, 19, 45.

57 Quoted in David Philip Miller, *The Life and Legend of James Watt: Collaboration, Natural Philosophy, and the Improvement of the Steam Engine* (Pittsburgh, Pa.: University of Pittsburgh Press, 2019), 176.

58 Crosbie Smith, *The Science of Energy: A Cultural History of Energy Physics in Victorian Britain* (Chicago: University of Chicago Press, 1998), 33. Max Weber's *Protestant Ethic and the Spirit of Capitalism* proposed more narrowly a sense of duty toward wealth, which he ascribed to anxiety over predestination. Weber, *The Protestant Ethic and the Spirit of Capitalism*, trans. by Talcott Parsons (New York: Scribner, 1930). Both ethic and spirit were broader in scope and the mechanism much more ingrained in Calvinist theology than Weber described.

59 Rudyard Kipling, "M'Andrew's Hymn," *The Seven Seas* (London: Methuen, 1896), 40, 31. *Institutio Christianae Religionis* was Calvin's classic theological work.

60 That religious belief correlated with certain economic attitudes seems incontrovertible. How and why it did so is complicated. See, for example, Barry Supple,

"The Nature of Enterprise," in *The Cambridge Economic History of Europe*, vol. 5, *The Economic Organisation of Early Modern Europe*, ed. by E. E. Rich and Charles Wilson (Cambridge: Cambridge University Press, 1977), 402–407.

61 Braudel, *Civilization and Capitalism*, 277–297, 353–385.

62 Richard Leslie Hills, *James* Watt, vol. 1, *His Time in Scotland, 1736–1774* (Ashbourne: Landmark, 2002), 150–154; Brian Watters, *Where Iron Runs Like Water!: A New History of Carron Iron Works, 1759–1982* (Edinburgh: John Donald, 1998), 1.

63 Hills, *Power from Steam*, 70.

64 A. E. Musson, "Industrial Motive Power in the United Kingdom, 1800–70," *Economic History Review*, n.s. 29(3) (1976): 417–418.

65 Sven Beckert, *Empire of Cotton: A Global History* (New York: Knopf, 2014), 19, 32.

66 Michael Herbert Fisher, *An Environmental History of India: From Earliest Times to the Twenty-First Century* (Cambridge: Cambridge University Press, 2018), 93–134.

67 Patrick Verley, *L'Échelle du Monde: Essai sur l'Industrialisation de l'Occident* (Paris: Gallimard, 1997), 160–164.

68 Broadberry and Gupta, "Lancashire, India, and Shifting Competitive Advantage," 298; Ian Wendt, "Writing the Rich Economic History of the South Asian Textile Industry: Spinners in Early Modern South India," in *Global Economic History Network Conference, Pune* (2005), http://www.lse.ac.uk/Economic-History/Assets/Documents/Research/GEHN/GEHNConferences/conf8/PUNEWendt.pdf, 8–9.

69 Stephen Broadberry and Bishnupriya Gupta, "Lancashire, India, and Shifting Competitive Advantage in Cotton Textiles, 1700–1850: The Neglected Role of Factor Prices," *Economic History Review*, n.s. 62(2) (2009): 288–291. On India and Europeans, see also Braudel, *Civilization and Capitalism*, vol. 3, 489–524.

70 John Lord, *Memoir of John Kay, of Bury: Inventor of the Fly-Shuttle* (Rochdale: Aldine Press, 1903). "Presbyterian": inferred, due to lack of specific information in surviving records; see 130 and passim.

71 R. S. Fitton, *The Arkwrights: Spinners of Fortune* (Manchester: Manchester University Press, 1989), 11–17; Presbyterian, 9.

72 Gilbert J. French, *The Life and Times of Samuel Crompton, Inventor of the Spinning Machine Called the Mule* (London: Simpkin, Marshall, 1859), 23–25, 120; Broadberry and Gupta, "Lancashire, India, and Shifting Competitive Advantage," 298.

73 Ibid., 284–286; Braudel, *Civilization and Capitalism*, vol. 3, 474.

74 Andreas Malm, *Fossil Capital: The Rise of Steam-Power and the Roots of Global*

Warming (London: Verso, 2016), 96–193. Malm argues for the superiority of water over steam power, discounts other problems with waterpower, and believes it was primarily owners' desire to control labor that led them to abandon waterpower.

75 Richard Leslie Hills, *James Watt*, vol. 3, *Triumph through Adversity, 1785–1819* (Ashbourne: Landmark, 2006), 63–68.

76 Musson, "Industrial Motive Power," 415–439; Boulton quoted on 429.

77 Beckert, *Empire of Cotton*, 136–174.

78 Henry Laufenburger, *Cours d'économie alsacienne* (Paris: Sirey, 1930–1932), vol. 1 97–114, vol. 2 133–235; Robert Fox, "Science, Industry, and the Social Order in Mulhouse, 1798–1871," *British Journal for the History of Science* 17(2) (1984): 127–168; Michael Stephen Smith, *The Emergence of Modern Business Enterprise in France, 1800–1930* (Cambridge: Harvard University Press, 2006), 131–144; Verley, *L'Échelle du Monde*, 168–170, 177; Serge Chassagne, *Le Coton et ses Patrons: France, 1760–1840* (Paris: Editions de l'École des Hautes Études en Sciences Sociales, 1991), 45–57, 75–80.

79 Barbara M. Tucker, "The Merchant, the Manufacturer, and the Factory Manager: The Case of Samuel Slater," *Business History Review* 55(3) (1981): 297–313.

80 See Chaim M. Rosenberg, *The Life and Times of Francis Cabot Lowell, 1775–1817* (Lanham, Md.: Lexington Books, 2011).

81 Thomas Dublin, *Women at Work: The Transformation of Work and Community in Lowell, Massachusetts, 1826–1860* (New York: Columbia University Press, 1979), 14–85; Theodore Steinberg, *Nature Incorporated: Industrialization and the Waters of New England* (Cambridge: Cambridge University Press, 1991), 21–96.

82 Ronald Bailey, "The Other Side of Slavery: Black Labor, Cotton, and Textile Industrialization in Great Britain and the United States," *Agricultural History* 68(2) (1994): 35–50; Ronald Bailey, "The Slave(ry) Trade and the Development of Capitalism in the United States: The Textile Industry in New England," *Social Science History* 14(3) (1990): 373–414; Richard L. Roberts, *Two Worlds of Cotton: Colonialism and the Regional Economy in the French Soudan, 1800–1946* (Stanford: Stanford University Press, 1996), 66.

83 Cowdrie, *This Land, This South*, 65–80; Mark Fiege, *The Republic of Nature: An Environmental History of the United States* (Seattle: University of Washington Press, 2012), 100–138; Erin Stewart Mauldin, *Unredeemed Land: An Environmental History of Civil War and Emancipation in the Cotton South* (New York: Oxford University Press, 2018).

84 Jonathan E. Robins, *Cotton and Race across the Atlantic: Britain, Africa, and America, 1900–1920* (Rochester, N.Y.: University of Rochester Press, 2016), 135–138.

85 A. Stowers, "Watermills, c.1500–c1850," in *A History of Technology*, vol. 4, *The Industrial Revolution c.1750 to c.1850*, ed. by Charles Singer, Eric John Holmyard, A. Rupert Hall, and Trevor Illtyd Williams (Oxford: Oxford University Press, 1958), 201.

86 Robert B. Gordon, "Cost and Use of Water Power during Industrialization in New England and Great Britain: A Geological Interpretation," *Economic History Review*, n.s. 36(2) (1983): 245.

87 See Theodore Steinberg, *Nature Incorporated: Industrialization and the Waters of New England* (Cambridge: Cambridge University Press, 1991).

4 *Age of Steam and Steel*

1 Adolph S. Cavallo, "To Set a Smart Board: Fashion as the Decisive Factor in the Development of the Scottish Linen Damask Industry," *Business History Review* 37(1/2) (1963): 49–58; David Nasaw, *Andrew Carnegie* (New York: Penguin Press, 2006), 1–23.

2 Alasdair Roberts, *America's First Great Depression: Economic Crisis and Political Disorder After the Panic of 1837* (Ithaca: Cornell University Press, 2012), 45; Marc-Antoine Longpré, John Stix, Cosima Burkert, Thor Hansteen, and Steffen Kutterolf, "Sulfur budget and global climate impact of the AD 1835 eruption of Cosigüina volcano, Nicaragua," *Geophysical Research Letters* 41(19) (2014): 6667–6675.

3 Stephen W. Campbell, "The Transatlantic Financial Crisis of 1837," in *Oxford Research Encyclopedia of Latin American History* (2017), ed. by William Beezley, https://oxfordre.com/latinamericanhistory/view/10.1093/acrefore/9780199366439.001.0001/acrefore-9780199366439-e-399.

4 Nasaw, *Carnegie*, 24–53.

5 Barry Eichengreen, *Globalizing Capital: A History of the International Monetary System*, 3rd edn. (Princeton: Princeton University Press, 2019), 5–40.

6 Benjamin Mountford and Stephen Tuffnell, "Seeking a Global History of Gold," 4–5, and Elliott West, "California, Coincidence, and Empire,"42–62, both in *A Global History of Gold Rushes*, ed. by Benjamin Mountford and Stephen Tuffnell (Oakland: University of California Press, 2018); H. Michell, "The Gold Standard in the Nineteenth Century," *Canadian Journal of Economics and Political Science / Revue Canadienne d'Economique et de Science Politique* 17(3) (1951): 369–376; David Vogel, *California Greenin': How the Golden State Became an Environmental Leader* (Princeton: Princeton University Press, 2018), 22–47.

7 David W. Miller, "Religious Commotions in the Scottish Diaspora: A Transatlantic Perspective on 'Evangelicalism' in a Mainline Denomination," in

Ulster Presbyterians in the Atlantic World: Religion, Politics and Identity, ed. by David A. Wilson and Mark G. Spencer (Dublin: Four Courts, 2006), 25.

8 Nasaw, *Carnegie*, 33–41, 54–57; Quentin R. Skrabec, *The Carnegie Boys: The Lieutenants of Andrew Carnegie That Changed America* (Jefferson, N.C.: McFarland, 2012), 32.

9 Andrew Carnegie, *Autobiography of Andrew Carnegie* (Boston: Houghton Mifflin, 1920), 40.

10 Justice Levi Woodbury, *Smith* v. *Downing*, in James D. Reid, *The Telegraph in America: Its Founders, Promoters and Noted Men* (New York: Derby Brothers, 1879), 110.

11 Richard B. Du Boff, "Business Demand and the Development of the Telegraph in the United States, 1844–1860," *Business History Review* 54(4) (1980): 459–479; James Schwoch, *Wired into Nature: The Telegraph and the North American Frontier* (Urbana: University of Illinois Press, 2018), 25–26.

12 Richard B. Du Boff, "The Telegraph in Nineteenth-Century America: Technology and Monopoly," *Comparative Studies in Society and History* 26(4) (1984): 572, 585.

13 Daniel Headrick, "A Double-Edged Sword: Communications and Imperial Control in British India," *Historical Social Research / Historische Sozialforschung* 35(1) (2010): 53–54; Daniel R. Headrick and Pascal Griset, "Submarine Telegraph Cables: Business and Politics, 1838–1939," *Business History Review* 75(3) (2001): 546–550; Claudia Steinwender, "Real Effects of Information Frictions: When the States and the Kingdom Became United," *American Economic Review* 108(3) (2018): 657–696.

14 Alfred D. Chandler, *The Visible Hand: The Managerial Revolution in American Business* (Cambridge: Harvard University Press, 1977), 197–200.

15 Sidney E. Morse to Samuel F. B. Morse, quoted in Reid, *Telegraph in America*, 89.

16 B. W. Clapp, *An Environmental History of Britain Since the Industrial Revolution* (London: Longman, 1994), 177–179.

17 Bode J. Morin, *The Legacy of American Copper Smelting: Industrial Heritage Versus Environmental Policy* (Knoxville: University of Tennessee Press, 2013), 7–14; Colin A. Russell and S. A. H. Wilmot, "Metal Extraction and Refining," in *Chemistry, Society and Environment: A New History of the British Chemical Industry*, ed. by Colin A. Russell (Cambridge: Royal Society of Chemistry, 2000), 295–300; quotation, *Report from the Select Committee of the House of Lords on Injury from Noxious Vapors, 1862*, in Russell and Wilmot, "Metal Extraction," 298.

18 Morin, *Legacy of American Copper Smelting*, 15–42; Duncan Maysilles, *Ducktown*

Smoke: The Fight Over One of the South's Greatest Environmental Disasters (Chapel Hill: University of North Carolina Press, 2011), 14–35.

19 Margaret Hindle Hazen and Robert M. Hazen, *Wealth Inexhaustible: A History of America's Mineral Industries to 1850* (New York: Van Nostrand Reinhold, 1985), 101.

20 Morin, *Legacy of American Copper Smelting*, 36.

21 Ibid., 16–36.

22 David Hochfelder, *The Telegraph in America, 1832–1920* (Baltimore: Johns Hopkins University Press, 2013), 186; William Henry Preece and James Sivewright, *Telegraphy* (London: Longman, Green, and Co., 1876), 8–38.

23 R. G. Coyle, *The Riches Beneath Our Feet: How Mining Shaped Britain* (New York: Oxford University Press, 2010), 74–92; Russell and Wilmot, "Metal Extraction and Refining," 308–309.

24 W. Ross Yates, "Samuel Wetherill, Joseph Wharton, and the Founding of the American Zinc Industry," *Pennsylvania Magazine of History and Biography* 98(4) (1974): 475.

25 Yates, "Samuel Wetherill," 479.

26 Cheryl M. Seeger, "History of Mining in the Southeast Missouri Lead District and Description of Mine Processes, Regulatory Controls, Environmental Effects, and Mine Facilities in the Viburnum Trend Subdistrict," in *Hydrologic Investigations Concerning Lead Mining Issues in Southeastern Missouri*, ed. by Michael J. Kleeschulte (Reston, Va.: U.S. Geological Survey, 2008); Jill McNew-Birren, *The Impacts of Lead Contamination on the Community of Herculaneum, Missouri* (PhD diss., Washington University, 2011), 97–98.

27 Henry Rowe Schoolcraft, *Journal of a Tour into the Interior of Missouri and Arkansaw, from Potosi, or Mire À Burton, in Missouri Territory, in a South-West Direction, Toward the Rocky Mountains, Performed in the Years 1818–1819* (London: Sir Richard Phillips, 1821), 4. See also Schoolcraft, *A View of the Lead Mines of Missouri; Including Some Observations on the Mineralogy, Geology, Geography, Antiquities, Soil Climate, Population, and Productions of Missouri and Arkansaw, and Other Sections of the Western Country* (New York: Wiley, 1819), 65.

28 B. H. Schockel, "Settlement and Development of the Lead and Zinc Mining Region of the Driftless Area with Special Emphasis Upon Jo Daviess County, Illinois," *The Mississippi Valley Historical Review* 4(2) (1917): 169–192; Robin Wall Kimmerer, "Vegetation Development on a Dated Series of Abandoned Lead and Zinc Mines in Southwestern Wisconsin," *American Midland Naturalist* 111(2) (1984): 332–341; Hazen and Hazen, *Wealth Inexhaustible*, 153–155.

29 Jie Ma et al., "Fractions and Colloidal Distribution of Arsenic Associated with Iron Oxide Minerals in Lead-Zinc Mine-Contaminated Soils: Comparison of

Tailings and Smelter Pollution," *Chemosphere* 227 (July 2019): 614–623; Bob Faust, "Lead in the Water: Power, Progressivism, and Resource Control in a Missouri Mining Community," *Agricultural History* 76(2) (2002): 413.

30 John Tully, "A Victorian Ecological Disaster: Imperialism, the Telegraph, and Gutta-Percha," *Journal of World History* 20(4) (2009): 559–579.

31 Cai Guise-Richardson, "Redefining Vulcanization: Charles Goodyear, Patents, and Industrial Control, 1834–1865," *Technology and Culture* 51(2) (2010): 357–387.

32 Warren Dean, *Brazil and the Struggle for Rubber: A Study in Environmental History* (Cambridge: Cambridge University Press, 1987), 4, 9, 38–41; John A. Tully, *The Devil's Milk: A Social History of Rubber* (New York: Monthly Review Press, 2011), 63–130; Corey Ross, *Ecology and Power in the Age of Empire: Europe and the Transformation of the Tropical World* (Oxford: Oxford University Press, 2017), 106–129; Gregg Mitman, *Empire of Rubber: Firestone's Scramble for Land and Power in Liberia* (New York: New Press, 2021).

33 Jean Gelman Taylor, *Global Indonesia* (Abingdon, Oxon: Routledge, 2013); G. Roger Knight, *Sugar, Steam and Steel: The Industrial Project in Colonial Java, 1830–1885* (Adelaide, Australia: University of Adelaide Press, 2014).

34 Barbara Watson Andaya and Leonard Y. Andaya, *A History of Malaysia*, 2nd edn. (Honolulu: University of Hawai'i Press, 2001), 133–204; Ross, *Ecology and Power in the Age of Empire*, 136–163.

35 Nasaw, *Carnegie*, 547–566, 603–604, 619–620, 648–649, 653–654.

36 Ernest Mahaim, "Les débuts de l'établissement John Cockerill à Seraing," *Vierteljahrschrift für Sozial- und Wirtschaftsgeschichte* 3(4) (1905): 627–648; Rainer Fremdling, "John Cockerill: Pionierunternehmer der Belgisch-niederländischen Industrialisierung," *Zeitschrift für Unternehmensgeschichte / Journal of Business History* 26(3) (1981): 179–193; A. Lecocq, *Description de l'établissement John Cockerill à Seraing* (Liège, Belgium: Gouchon, 1854), 13–18.

37 George Rogers Taylor, *The Transportation Revolution, 1815–1860* (White Plains, N.Y.: M. E. Sharpe, 1951), 74–90.

38 Ibid., 56–73.

39 Peter A. Shulman, *Coal and Empire: The Birth of Energy Security in Industrial America* (Baltimore: Johns Hopkins University Press, 2015), 125–163.

40 Taylor, *Transportation Revolution*, 396–398.

41 Nasaw, *Carnegie*, 54–65.

42 Chandler, *Visible Hand*, 79–133.

43 Nasaw, *Carnegie*, 66–88.

44 Ibid., 89–114.

45 Rainer Fremdling, "Transfer Patterns of British Technology to the Continent:

The Case of the Iron Industry," *European Review of Economic History* 4(2) (2000): 196; Morgan Kelly, Joel Mokyr, and Cormac Ó Gráda, "Could Artisans Have Caused the Industrial Revolution?," in *Reinventing the Economic History of Industrialisation*, ed. by Kristine Bruland, Anne Gerritsen, Pat Hudson, and Giorgio Riello (Montreal: McGill–Queen's University Press, 2020), 25–43.

46 Norbert C. Soldon, *John Wilkinson, 1728–1808: English Ironmaster and Inventor* (Lewiston, N.Y.: Edwin Mellen Press, 1998), 24.

47 Soldon, *John Wilkinson*, 17–25; Tom Williamson, Gerry Barnes, and Toby Pillatt, *Trees in England: Management and Disease Since 1600* (Hatfield: University of Hertfordshire Press, 2017), 132–162.

48 William Stanley Jevons, *The Coal Question*, 2nd edn. (London: Macmillan, 1866), 312–324; Chris Evans, Owen Jackson, and Göran Rydén, "Baltic Iron and the British Iron Industry in the Eighteenth Century," *Economic History Review*, 55(4) (2002): 645.

49 Alfred D. Chandler, "Anthracite Coal and the Beginnings of the Industrial Revolution in the United States," *Business History Review* 46(2) (1972): 141–148.

50 Christopher F. Jones, *Routes of Power: Energy and Modern America* (Cambridge: Harvard University Press, 2014), 23–87.

51 Lewis, W. David. "The Early History of the Lackawanna Iron and Coal Company: A Study in Technological Adaptation," *Pennsylvania Magazine of History and Biography* 96(4) (1972): 424–468; Chandler, "Anthracite Coal," 142–149, 156–158, 176–181; quotation on 149; Jones, *Routes of Power* 23–88.

52 James Parton, "Pittsburgh," *Atlantic Monthly*, January 21, 1868, 21.

53 Hazen and Hazen, *Wealth Inexhaustible*, 206–211; Joel A. Tarr and Karen Clay, "Pittsburgh as an Energy Capital: Perspectives on Coal and Natural Gas Transitions and the Environment," in *Energy Capitals: Local Impact, Global Influence*, ed. by Joseph A. Pratt, Martin V. Melosi, and Kathleen A. Brosnan (Pittsburgh, Pa.: University of Pittsburgh Press, 2014), 5–11.

54 Chandler, *Visible Hand*, 91–93.

55 Nasaw, *Carnegie*, 105–136.

56 Ibid., 137–144.

57 Thomas J. Misa, *A Nation of Steel: The Making of Modern America, 1865–1925* (Baltimore: Johns Hopkins University Press, 1995), 1–15; Geoffrey Tweedale, "Metallurgy and Technological Change: A Case Study of Sheffield Specialty Steel and America, 1830–1930," *Technology and Culture* 27(2) (1986): 189–222.

58 Misa, *Nation of Steel*, 15–21.

59 Krupp'sche Gußstahlfabrik, *Krupp 1812–1912: zum 100 jährigen Bestehen der Firma Krupp und der Gußstahlfabrik zu Essen-Ruhr* (Essen, Germany: Krupp, 1912), 268.

60 Harold James, *Krupp: A History of the Legendary German Firm* (Princeton: Princeton University Press, 2012), 15–51.

61 Jim Clifford, "Mapping Supply Chains for 19th Century Leather" (August 12, 2014), http://www.jimclifford.ca/2014/08/12/mapping-supply-chains-for-19th-century-leather/.

62 M. Scott Taylor, "Buffalo Hunt: International Trade and the Virtual Extinction of the North American Bison," *American Economic Review* 101(7) (2011): 3162–3171, 3189; see also Dan Flores, "Bison Ecology and Bison Diplomacy: The Southern Plains from 1800 to 1850," *Journal of American History* 78(2) (1991): 465–485; Andrew C. Isenberg, *The Destruction of the Bison: An Environmental History, 1750–1920* (Cambridge: Cambridge University Press, 2000); and Jennifer Hansen, "A Tanner's View of the Bison Hunt: Global Tanning and Industrial Leather," in *Bison and People on the North American Great Plains: A Deep Environmental History*, ed. by Geoff Cunfer and W. A. Waiser (College Station: Texas A&M University Press, 2016), 227–244.

63 William Cronon, *Nature's Metropolis: Chicago and the Great West* (New York: Norton, 1991), 218–259. On environmental impacts of transcontinental railroads, see also Mark Fiege, *The Republic of Nature: An Environmental History of the United States* (Seattle: University of Washington Press, 2012), 228–280.

64 Cronon, *Nature's Metropolis*, 97–147. See also Richard White, *Railroaded: The Transcontinentals and the Making of Modern America* (New York: Norton, 2011).

65 Cronon, *Nature's Metropolis*, 313–318, 148–206, 324–340, 263–295.

66 Sergei Witte, *Vorlesungen über Volks- und Staatswirtschaft*, quoted in Steven G. Marks, *Road to Power: The Trans-Siberian Railroad and the Colonization of Asian Russia, 1850–1917* (Ithaca, N.Y.: Cornell University Press, 1991), 126.

67 Marks, *Road to Power*, 125–169, 196–219.

68 Nasaw, *Carnegie*, 144–150.

69 Misa, *Nation of Steel*, 21–28.

70 Nasaw, *Carnegie*, 150–183.

71 Misa, *Nation of Steel*, 45–89; Ken Kobus, *City of Steel: How Pittsburgh Became the World's Steelmaking Capital during the Carnegie Era* (Lanham, Md.: Rowman & Littlefield, 2015), 149–256; Nasaw, *Carnegie*, 247, 361–363, 377–378, 405, 469.

72 Robert Marks, *The Origins of the Modern World: A Global and Environmental Narrative from the Fifteenth to the Twenty-First Century*, 3rd edn. (Lanham, Md.: Rowman & Littlefield, 2015), 144; David S. Landes, *The Unbound Prometheus: Technological Change and Industrial Development in Western Europe from 1750 to the Present* (Cambridge: Cambridge University Press, 1969), 6; Michael Stephen Smith, *The Emergence of Modern Business Enterprise in France, 1800–1930* (Cambridge: Harvard University Press, 2006), 137–142.

73 Herbert Kisch, *From Domestic Manufacture to Industrial Revolution: The Case of the Rhineland Textile Districts* (New York: Oxford University Press, 1989); Elaine Glovka Spencer, "Rulers of the Ruhr: Leadership and Authority in German Big Business before 1914," *Business History Review* 53(1) (1979): 40–64; Jonathan Sperber, "The Shaping of Political Catholicism in the Ruhr Basin, 1848–1881," *Central European History* 16(4) (1983): 347–367; Marks, *The Origins of the Modern World*, 145–146; Mark Cioc, *The Rhine: An Eco-Biography, 1815–2000* (Seattle: University of Washington Press, 2002), 47–107.

74 Marks, *The Origins of the Modern World*, 146–147; Boris Kagarlitsky, *Empire of the Periphery: Russia and the World System*, trans. by Renfrey Clarke (London: Pluto Press, 2008), 223–282; Roman Adrian Cybriwsky, *Along Ukraine's River: A Social and Environmental History of the Dnipro (Dnieper)* (Budapest: Central European University Press, 2018), 64–65, 71.

75 Nasaw, *Carnegie*, 513–523.

76 Duane A. Smith, *Mining America: The Industry and the Environment, 1800–1980* (Lawrence: University Press of Kansas, 1987), 86–88, 110–111, 114–116.

77 Klaus H. Wolff, "Textile Bleaching and the Birth of the Chemical Industry," *Business History Review* 48(2) (1974): 144–150.

78 Ibid., 150–161.

79 Christian Simon, "The Rise of the Swiss Chemical Industry Reconsidered," in *The Chemical Industry in Europe, 1850–1914: Industrial Growth, Pollution, and Professionalization*, ed. by Ernst Homburg, Anthony S. Travis, and Harm G. Schröter (Dordrecht, Netherlands: Kluwer Academic, 1998), 9–28.

80 Arne Andersen, "Pollution and the Chemical Industry: The Case of the German Dye Industry," in *The Chemical Industry in Europe*, 183–186.

81 Carnegie, quoted in Nasaw, *Andrew Carnegie*, 113–114.

82 Nasaw, *Andrew Carnegie*, 345–354.

83 Ibid., 589–614, 712–717.

84 Ibid., 543–566, 641–711, 724–756, 770–801; see also Misa, *Nation of Steel*, 270–278.

5 *Conserving Resources*

1 W. Stanley Jevons, *The Coal Question*, 2nd edn. (London: Macmillan, 1866), 371.

2 George P. Marsh, *Man and Nature, or, Physical Geography As Modified by Human Action* (New York: C. Scribner, 1864), 43.

3 Ibid., 48.

4 Ibid., 55.

5 Ibid., 326.

6 Ibid., 122, 123.
7 Ibid., 37n, 113.
8 Jevons, *Coal Question*, 1.
9 Ibid., 101–137.
10 Ibid., xxvi.
11 Ibid., 376.
12 Nuno Luis Madureira, "The Anxiety of Abundance: William Stanley Jevons and Coal Scarcity in the Nineteenth Century," *Environment and History* 18(3) (2012): 395–421; Rosamond Könekamp, "Biographical Introduction," in William Stanley Jevons, *Papers and Correspondence*, vol. 1, *Biography and Personal Journal*, ed. by R. D. Collison Black and Rosamond Könekamp (London: Macmillan, 1972), 44–45.
13 David Lowenthal, *George Perkins Marsh: Prophet of Conservation* (Seattle: University of Washington Press, 2000), 1–21.
14 Lowenthal, *Marsh*, 21–369. See also George P. Marsh, *Address Delivered Before the Agricultural Society of Rutland County, Sept. 30, 1847* (Rutland, Vt., 1848).
15 Jevons, *Letters and Journal of W. Stanley Jevons*, ed. by Harriet A. Jevons (London: Macmillan, 1886) (quotations on 144, 146); Könekamp, "Biographical Introduction," 1–52; Bert Mosselmans, *William Stanley Jevons and the Cutting Edge of Economics* (London: Routledge, 2007), 1–6.
16 Lowenthal, *Marsh*, 7–11, 20–21, 63–67, 375–377; quotation, 377.
17 Jevons, *Papers*, 1, 2–3, 29, 52; Mosselmans, *William Stanley Jevons*, 63–82.
18 See David D. Hall, *A Reforming People: Puritanism and the Transformation of Public Life in New England* (New York: Knopf, 2011); Brian Donahue, *The Great Meadow: Farmers and the Land in Colonial Concord* (New Haven: Yale University Press, 2004).
19 Mark Stoll, "The Other Scientific Revolution: Calvinist Scientists and the Origins of Ecology," in *After the Death of Nature: Carolyn Merchant and the Future of Human-Nature Relations*, ed. by Kenneth Worthy, Elizabeth Allison, and Whitney Bauman (New York: Routledge, 2018).
20 See Susan Elizabeth Schreiner, *The Theater of His Glory: Nature and the Natural Order in the Thought of John Calvin* (Durham, N.C.: Labyrinth Press, 1991); and Mark R. Stoll, *Inherit the Holy Mountain: Religion and the Rise of American Environmentalism* (Oxford: Oxford University Press, 2015), 10–76.
21 Marsh, *Man and Nature*, 35–36.
22 Ibid., 54n.
23 Ibid., 8.
24 Ibid., 35.
25 Ibid., 35.

26 Ibid., 303.

27 Mosselmans, *William Stanley Jevons*, 3–6; H. L. Short, "Presbyterians under a New Name," in C. Gordon Bolam, Jeremy Goring, H. L. Short, and Roger Thomas, *The English Presbyterians: From Elizabethan Puritanism to Modern Unitarianism* (London: Allen & Unwin, 1968).

28 Jevons, *Coal Question*, xxiii, xxv, xxvi.

29 Henri Gourdin, *Olivier de Serres: "Science, expérience, diligence" en agriculture au temps de Henri IV* (Arles, France: Actes Sud, 2001); Jean Boulaine, "Innovations agronomiques d'Olivier de Serres," *Bulletin de l'Association d'étude sur l'humanisme, la réforme et la renaissance* 50 (2000): 11–19; Danièle Duport, "La 'science' d'Olivier de Serres et la connaissance du 'naturel,'" *Bulletin de l'Association d'étude sur l'humanisme, la réforme et la renaissance* 50 (2000): 185–95; Denise le Dantec and Jean-Pierre le Dantec, *Reading the French Garden: Story and History*, trans. by Jessica Levine (Cambridge: MIT Press, 1990), 64–74; Henry Heller, *The Conquest of Poverty: The Calvinist Revolt in Sixteenth-Century France* (Leiden: Brill, 1986); Mark Stoll, "'Sagacious' Bernard Palissy: Pinchot, Marsh, and the Connecticut Origins of American Conservation," *Environmental History* 16(1) (2011): 4–37.

30 Samuel Gardiner, *Doomes-Day Booke: or, an Alarum for Atheistes, a Watchword for Worldlinges, a Caueat for Christians* (London, 1606), 107. See Nicholas Tyacke, *Anti-Calvinists: The Rise of English Arminianism, c.1590–1640* (Oxford: Clarendon Press, 1987), 254–256.

31 Paul Warde, *The Invention of Sustainability: Nature and Destiny, c.1500–1870* (Cambridge: Cambridge University Press, 2018), 102–143; Warde, "The Idea of Improvement, c.1520–1770," in *Custom, Improvement and the Landscape in Early Modern Britain*, ed. by R. W. Hoyle (Farnham, Surrey: Ashgate, 2011), 127–148; Stoll, "'Sagacious' Palissy," 18, 19–20.

32 Patricia James, *Population Malthus: His Life and Times* (London: Routledge & Kegan Paul, 1979), 19–23.

33 Vernon W. Ruttan, *Agricultural Research Policy* (Minneapolis: University of Minnesota Press, 1982), 72–73.

34 Gregory T. Cushman, *Guano and the Opening of the Pacific World: A Global Ecological History* (Cambridge: Cambridge University Press, 2013), 23–74; Warde, *Invention of Sustainability*, 297–307; Steven Stoll, *Larding the Lean Earth: Soil and Society in Nineteenth-Century America* (New York: Hill and Wang, 2002), 187–194; Deborah Fitzgerald, *Every Farm a Factory: The Industrial Ideal in American Agriculture* (New Haven: Yale University Press, 2003).

35 Mark Overton, *Agricultural Revolution in England: The Transformation of the Agrarian Economy, 1500–1850* (Cambridge: Cambridge University Press, 1996),

193–207; J. D. Chambers and G. E. Mingay, *The Agricultural Revolution, 1750–1880* (New York: Schocken Books, 1966); Ruttan, *Agricultural Research Policy*, 67–68; Richard Perren, *Agriculture in Depression, 1870–1940* (Cambridge: Cambridge University Press, 1995), 1–16.

36 Erin Stewart Mauldin, *Unredeemed Land: An Environmental History of Civil War and Emancipation in the Cotton South* (New York: Oxford University Press, 2018).

37 Perren, *Agriculture in Depression*, 17–30.

38 Margaret W. Rossiter, *The Emergence of Agricultural Science: Justus Liebig and the Americans, 1840–1880* (New Haven: Yale University Press, 1975); Ruttan, *Agricultural Research Policy*, 76–77.

39 Conrad D. Totman, *The Green Archipelago: Forestry in Preindustrial Japan* (Berkeley: University of California Press, 1989); Mark Elvin, *The Retreat of the Elephants: An Environmental History of China* (New Haven: Yale University Press, 2004), 9–39; Meng Zhang, *Timber and Forestry in Qing China: Sustaining the Market* (Seattle: University of Washington Press, 2021); Chetan Singh, "Forests, Pastoralists and Agrarian Society in Mughal India," in *Nature, Culture, Imperialism: Essays on the Environmental History of South Asia*, ed. by David Arnold and Ramachandra Guha (Delhi, India: Oxford University Press, 1995); J. R. McNeill, "Woods and Warfare in World History," *Environmental History* 9(3) (2004): 405; Madhav Gadgil and Ramachandra Guha, *This Fissured Land: An Ecological History of India* (Berkeley: University of California Press, 1993), 85–90, 106–108.

40 Clarence J. Glacken, *Traces on the Rhodian Shore: Nature and Culture in Western Thought from Ancient Times to the End of the Eighteenth Century* (Berkeley: University of California Press, 1967), 126–137, 311–351; Warde, *Invention of Sustainability*, 58–101.

41 See Karl Richard Appuhn, *A Forest on the Sea: Environmental Expertise in Renaissance Venice* (Baltimore: Johns Hopkins University Press, 2009).

42 Alan Mikhail, *Under Osman's Tree: The Ottoman Empire, Egypt, and Environmental History* (Chicago: University of Chicago Press, 2017), 153–168.

43 John T. Wing, "Keeping Spain Afloat: State Forestry and Imperial Defense in the Sixteenth Century," *Environmental History* 17(1) (January 2012): 116–145.

44 McNeill, "Woods and Warfare," 396–398.

45 Warde, *Invention of Sustainability*, 61–65, 91–92; Greg Barton, *Empire Forestry and the Origins of Environmentalism* (Cambridge: Cambridge University Press, 2002), 12.

46 Warde, *Invention of Sustainability*, 88–90; Barton, *Empire Forestry*, 13–15.

47 Warde, *Invention of Sustainability*, 153–159, 165–171, 175–182, 200–227; Jan

Oosthoek, "Worlds Apart? The Scottish Forestry Tradition and the Development of Forestry in India," *Journal of Irish and Scottish Studies* 3(1) (2010): 71; S. Ravi Rajan, *Modernizing Nature: Forestry and Imperial Eco-Development 1800–1950* (Oxford: Oxford University Press, 2006), 21–54.

48 John M. MacKenzie, "Scots and the Environment of Empire," in *Scotland and the British Empire*, ed. by John M. MacKenzie and T. M Devine (Oxford: Oxford University Press, 2011), 150–160, 162; see also V. H. Heywood, "The Changing Rôle of the Botanic Gardens," in *Botanic Gardens and the World Conservation Strategy: Proceedings of an International Conference, 26–30 November 1985, Held at Las Palmas De Gran Canaria*, ed. by D. Bramwell, O. Hamann, V. Heywood, and H. Synge (London: Academic Press, 1987), 3–18.

49 Lowenthal, *George Perkins Marsh*, 304, 509n33. On Grove's dismissal of Marsh's influence, see 421–422. Overall, Grove claims too much significance for islands and for indigenous Indian notions.

50 Oosthoek, "Worlds Apart?," 72–78; Grove, *Green Imperialism*, 380–473. See also Barton, *Empire Forestry*, 38–129.

51 Arthur A. Ekirch, "Franklin B. Hough: First Citizen of the Adirondacks," *Environmental Review* 7(3) (1983): 271–274.

52 Edna L. Jacobsen, "Franklin B. Hough: A Pioneer in Scientific Forestry in America," *New York History* 15(3) (1934): 311–325.

53 Char Miller, "Amateur Hour: Nathaniel H. Egleston and Professional Forestry in Post-Civil War America," *Forest History Today* (Spring/Fall 2005): 20–26.

54 Char Miller, *Gifford Pinchot and the Making of Modern Environmentalism* (Washington, D.C.: Island Press, 2001).

55 Ibid., 77–176; Barton, *Empire Forestry*, 130–143; Uwe E. Schmidt, "German Impact and Influences on American Forestry Until World War II," *Journal of Forestry* 107(3) (Apr. 2009): 139–145.

56 Thomas R. Dunlap, *Saving America's Wildlife* (Princeton: Princeton University Press, 1988), 1–61; Jeff Schauer, *Wildlife Between Empire and Nation in Twentieth-Century Africa* (Cham, Switzerland: Palgrave Macmillan, 2018), 17–70.

57 William M. Cavert, *The Smoke of London: Energy and Environment in the Early Modern City* (Cambridge: Cambridge University Press, 2016).

58 Harold L. Platt, *Shock Cities: The Environmental Transformation and Reform of Manchester and Chicago* (Chicago: University of Chicago Press, 2005), 24–48.

59 Javier Abellán, "Water Supply and Sanitation Services in Modern Europe: Developments in 19th–20th Centuries" (paper presented at the 12th International Congress of the Spanish Association of Economic History, University of Salamanca, September 2017).

60 Christopher Hamlin, *Public Health and Social Justice in the Age of Chadwick: Britain, 1800–1854* (Cambridge: Cambridge University Press, 1998), 156–187.

61 Martin V. Melosi, *The Sanitary City: Urban Infrastructure in America from Colonial Times to the Present* (Baltimore: Johns Hopkins University Press, 2000), 43–174. For Manchester, see Platt, *Shock Cities.*

62 See Leo Marx, *The Machine in the Garden: Technology and the Pastoral Ideal in America* (New York: Oxford University Press, 1964); Eric Purchase, *Out of Nowhere: Disaster and Tourism in the White Mountains* (Baltimore: Johns Hopkins University Press, 1999).

63 Jere Stuart French, "The First 'People's Park' Movement," *Landscape Architecture* 62(1) (1971): 25–29; Hilary A. Taylor, "Urban Public Parks, 1840–1900: Design and Meaning," *Garden History* 23(2) (1995): 201–203; John W. Henneberger, "Origins of Fully Funded Public Parks," *George Wright Forum* 19(2) (2002): 13–17; Brent Elliott, "Loudon, John Claudius (1783–1843), Landscape Gardener and Horticultural Writer," *Oxford Dictionary of National Biography*, September 23, 2004, https://www-oxforddnb-com.lib-e2.lib.ttu.edu/view/10.1093/ref:odnb/9780198614128.001.0001/odnb-9780198614128-e-17031.

64 Ophélie Siméon, *Robert Owen's Experiment at New Lanark: From Paternalism to Socialism* (London: Palgrave Macmillan, 2017).

65 Anthony Sutcliffe, *Towards the Planned City: Germany, Britain, the United States, and France, 1780–1914* (New York: St. Martin's Press, 1981), 146–148, 151, 191; John S. Garner, *The Model Company Town: Urban Design through Private Enterprise in Nineteenth-Century New England* (Amherst: University of Massachusetts Press, 1984), 111–113; Henry Roberts, *The Dwellings of the Labouring Classes, Their Arrangement and Construction* (London: Society for Improving the Condition of the Labouring Classes, 1850), 24; Eliseu Gonçalves and Rui J. G. Ramos, "Primeiras Propostas de Habitação Operária no Porto: A Casa Unifamiliar, o Carré Mulhousien e a Cité-jardin," *Ciudades (Valladolid, Spain)* 19 (2016): 77–82.

66 Hardy Green, *The Company Town: The Industrial Edens and Satanic Mills That Shaped the American Economy* (Boulder, Colo.: Basic Books, 2010), 57–89; Thomas G. Andrews, *Killing for Coal: America's Deadliest Labor War* (Cambridge: Harvard University Press, 2008).

67 Sarah Bilston, *The Promise of the Suburbs: A Victorian History in Literature and Culture* (New Haven: Yale University Press, 2019), 20–36.

68 Stanley Buder, *Visionaries and Planners: The Garden City Movement and the Modern Community* (Oxford: Oxford University Press, 1990); Michael Bally and Stephen Marshall, "Centenary Paper: The Evolution of Cities: Geddes, Abercrombie and the New Physicalism," *Town Planning Review* 80(6) (2009):

551–574; Helen Elizabeth Meller, *Patrick Geddes: Social Evolutionist and City Planner* (London: Routledge, 1989).

69 Marsh, *Man and Nature*, 235.

70 Terrie, Philip G., *Forever Wild: Environmental Aesthetics and the Adirondack Forest Preserve* (Philadelphia: Temple University Press, 1985), 92–108.

71 Frank Uekötter, *The Age of Smoke: Environmental Policy in Germany and the United States, 1880–197* (Pittsburgh, Pa.: University of Pittsburgh Press, 2009), 20–66; Andrew Lees, *Cities, Sin, and Social Reform in Imperial Germany* (Ann Arbor: University of Michigan Press, 2002), 45–48.

72 Antoine Missemer, *Les Économistes et la fin des énergies fossiles (1865–1931)* (Paris: Classiques Garnier, 2017), 49–51.

73 Robert Morse Crunden, *Ministers of Reform: The Progressives' Achievement in American Civilization, 1889–1920* (New York: Basic Books, 1982); Stoll, *Inherit the Holy Mountain*, 137–171, appendix.

74 See Gregg Mitman, *The State of Nature: Ecology, Community, and American Social Thought, 1900–1950* (Chicago: University of Chicago Press, 1992); Ronald C. Tobey, *Saving the Prairies: The Life Cycle of the Founding School of American Plant Ecology, 1895–1955* (Berkeley: University of California Press, 1981); Joel B. Hagen, *An Entangled Bank: The Origins of Ecosystem Ecology* (New Brunswick, N.J.: Rutgers University Press, 1992).

75 Douglas Brinkley, *The Wilderness Warrior: Theodore Roosevelt and the Crusade for America* (New York: HarperCollins, 2009); Martin W. Holdgate, *The Green Web: A Union for World Conservation* (Cambridge: Earthscan, 1999), 11.

76 Holdgate, *Green Web*, 6–13.

77 Stephen R. Fox, *The American Conservation Movement: John Muir and his Legacy* (Madison: University of Wisconsin Press, 1985), 148–182.

78 Douglas Brinkley, *Rightful Heritage: Franklin D. Roosevelt and the Land of America* (New York: Harper, 2016); Neil M. Maher, *Nature's New Deal: The Civilian Conservation Corps and the Roots of the American Environmental Movement* (New York: Oxford University Press, 2008); Donald Worster, *Dust Bowl: The Southern Plains in the 1930s* (New York: Oxford University Press, 1979).

79 Holdgate, *Green Web*, 11–38.

80 B. W. Clapp, *An Environmental History of Britain Since the Industrial Revolution* (London: Longman, 1994), 16; David Stradling, *Smokestacks and Progressives: Environmentalists, Engineers and Air Quality in America, 1881–1951* (Baltimore: Johns Hopkins University Press, 1999), 13 Table 1, 187 Table 4.

6 Buy Now — Pay Later

1 Alfred D. Chandler, *Giant Enterprise: Ford, General Motors, and the Automobile Industry: Sources and Readings* (New York: Harcourt Brace & World, 1964), 3.

2 Richard P. Scharchburg, *Carriages Without Horses: J. Frank Duryea and the Birth of the American Automobile Industry* (Warrendale, Pa.: Society of Automotive Engineers, 1993), 89–100, 141–160.

3 Peter Dauvergne, *The Shadows of Consumption: Consequences for the Global Environment* (Cambridge: MIT Press, 2008), 37–38; William Pelfrey, *Billy, Alfred, and General Motors: The Story of Two Unique Men, a Legendary Company, and a Remarkable Time in American History* (New York: AMACOM, 2006), 51–75; Sally H. Clarke, *Trust and Power: Consumers, the Modern Corporation, and the Making of the United States Automobile Market* (Cambridge: Cambridge University Press, 2007), 3.

4 Pelfrey, *Billy, Alfred, and General Motors*, 43–49.

5 Ibid., 17–36, 93–236; Lawrence R. Gustin, *Billy Durant: Creator of General Motors*, 3rd edn. (Ann Arbor: University of Michigan Press, 2008), 26–222. Unlike Ford or Sloan, Durant was a Presbyterian (33).

6 Stuart W. Leslie, *Boss Kettering* (New York: Columbia University Press, 1983), 90–97.

7 David R. Farber, *Sloan Rules: Alfred P. Sloan and the Triumph of General Motors* (Chicago: University of Chicago Press, 2002), 1–105; Pelfrey, *Billy, Alfred, and General Motors*, 237–272.

8 Robert Sobel, *The Great Bull Market: Wall Street in the 1920s* (New York: Norton, 1968), 45.

9 Frank Trentmann, *Empire of Things: How We Became a World of Consumers, from the Fifteenth Century to the Twenty-First* (New York: HarperCollins, 2016), 21–118; Peter N. Stearns, *Consumerism in World History: The Global Transformation of Desire*, 2nd edn. (London: Routledge, 2006), 1–43.

10 Patrick Verley, *L'Échelle du Monde: Essai sur l'industrialisation de l'Occident* (Paris: Gallimard, 1997), 136–139.

11 William Leach, *Land of Desire: Merchants, Power, and the Rise of a New American Culture* (New York: Pantheon Books, 1993), 3–150; Boris Emmet and John E. Jeuck, *Catalogues and Counters: A History of Sears, Roebuck and Company* (Chicago: University of Chicago Press, 1950), 9–99.

12 Susie Pak, *Gentlemen Bankers: The World of J. P. Morgan* (Cambridge: Harvard University Press, 2014), 13; Kenneth Warren, *Big Steel: The First Century of the United States Steel Corporation, 1901–2001* (Pittsburgh, Pa.: University of Pittsburgh Press, 2001), 7–21; Robert Hessen, *Steel Titan: The Life of Charles M. Schwab* (New York: Oxford University Press, 1975), 111–122; David Nasaw,

Andrew Carnegie (New York: Penguin Press, 2006), 580–588; Thomas Philippon, "Has the US Finance Industry Become Less Efficient? On the Theory and Measurement of Financial Intermediation," *American Economic Review* 105(4) (2015): 1408–38. See Gerald F. Davis, *Managed by the Markets: How Finance Reshaped America* (Oxford: Oxford University Press, 2009). Durant's financial activities are a major theme in Pelfrey, *Billy, Alfred, and General Motors.*

13 Emmet and Jeuck, *Catalogues and Counters*, 338–357; Leach, *Land of Desire*, 263–285.

14 Rowena Olegario, *The Engine of Enterprise: Credit in America* (Cambridge: Harvard University Press, 2016), 49–50.

15 Norton Reamer and Jesse Downing, *Investment: A History* (New York: Columbia University Press, 2016), 62–98.

16 Peter Lester Payne, *British Entrepreneurship in the Nineteenth Century*, 2nd edn. (Houndmills, Basingstoke, Hampshire: Macmillan Education, 1988), 55–57; Youssef Cassis, "British Finance: Success and Controversy," in *Capitalism in a Mature Economy: Financial Institutions, Capital Exports and British Industry, 1870–1939*, ed. by J. J. Van Helten and Youssef Cassis (Aldershot, Hants, England: Elgar, 1990), 1–22.

17 Olegario, *Engine of Enterprise*, 67.

18 Ibid., 108.

19 Ibid., 97.

20 Ibid., 133–136.

21 Sobel, *The Great Bull Market*, 45.

22 Olegario, *Engine of Enterprise*, 132–133; Susan Porter Benson, "Gender, Generation, and Consumption in the United States: Working-Class Families in the Interwar Period," in *Getting and Spending: European and American Consumer Societies in the Twentieth Century*, ed. by Susan Strasser, Charles McGovern, and Matthias Judt (Cambridge: Cambridge University Press, 1998), 238–239.

23 Leach, *Land of Desire*, 124–125; Olegario, *Engine of Enterprise*, 143–144.

24 Daniel Yergin, *The Prize: The Epic Quest for Oil, Money, and Power* (New York: Simon & Schuster, 1991), 543–546; Peter A. Shulman, *Coal and Empire: The Birth of Energy Security in Industrial America* (Baltimore: Johns Hopkins University Press, 2015), 164–213; Timothy Mitchell, *Carbon Democracy: Political Power in the Age of Oil* (London: Verso, 2011), 12–42.

25 Christian Pfister, "The 'Syndrome of the 1950s' in Switzerland: Cheap Energy, Mass Consumption, and the Environment," in *Getting and Spending*, ed. by Strasser, McGovern, and Judt, 359–377.

26 Vaclav Smil, *Energy and Civilization: A History* (Cambridge: MIT Press, 2017), 245–247.

27 Nasaw, *Andrew Carnegie*, 74–79, 141.
28 Darren Dochuk, *Anointed with Oil: How Christianity and Crude Made Modern America* (New York: Basic Books, 2019), 21–138; Yergin, *The Prize*, 19–95.
29 Yergin, *The Prize*, 111.
30 Smil, *Energy and Civilization*, 259, 261–262.
31 Vaclav Smil, *Creating the Twentieth Century: Technical Innovations of 1867–1914 and Their Lasting Impact* (Oxford: Oxford University Press, 2005), 33–97.
32 Sobel, *Great Bull Market*, 36.
33 Trentmann, *Empire of Things*, 248–249.
34 See James C. Williams, *Energy and the Making of Modern California* (Akron: University of Akron Press, 1997).
35 Kathryn Steen, *The American Synthetic Organic Chemicals Industry: War and Politics, 1910–1930* (Chapel Hill: University of North Carolina Press, 2014), 156–171, 237–238, 255, 262–267, 280.
36 Ibid., 282–285; Jeffrey L. Meikle, "Into the Fourth Kingdom: Representations of Plastic Materials, 1920–1950," *Journal of Design History* 5(3) (1992): 173–182. See also Stephen Fenichell, *Plastic: The Making of a Synthetic Century* (New York: HarperBusiness, 1996), 1–223.
37 Steen, *American Synthetic Organic Chemicals Industry*, 263–264; John Robert McNeill, *Something New Under the Sun: An Environmental History of the Twentieth-Century World* (New York: Norton, 2000), 111–113.
38 Giles Slade, *Made to Break: Technology and Obsolescence in America* (Cambridge: Harvard University Press, 2006), 29–55.
39 Leonard S. Reich, "Lighting the Path to Profit: GE's Control of the Electric Lamp Industry, 1892–1941," *Business History Review* 66(2) (1992): 305–334; Slade, *Made to Break*, 31; Wyatt C. Wells, *Antitrust and the Formation of the Postwar World* (New York: Columbia University Press, 2002), 19–22.
40 Trentmann, *Empire of Things*, 280–283.
41 Slade, *Made to Break*, 78–81.
42 Ibid., 15–17; Randal C. Picker, "The Razors-and-Blades Myth(s)," *University of Chicago Law Review* 78(1) (2011): 225–255; Gordon McKibben, *Cutting Edge: Gillette's Journey to Global Leadership* (Boston: Harvard Business School Press, 1998), 5–22.
43 Herbert Hoover, *Remarks of President Hoover at the Dinner of the Association of National Advertisers: Washington, D.C., November 10, 1930* (Washington: U.S. Government Printing Office, 1930), 1.
44 This is Jacques Ellul's concept of "sociological propaganda." See Ellul, *Propaganda: The Formation of Men's Attitudes*, trans. by Konrad Kellen and Jean Lerner (New York: Knopf, 1965), 62–70.

45 Daniel Navon, "Truth in Advertising: Rationalizing Ads and Knowing Consumers in the Early Twentieth-Century United States," *Theory and Society* 46(2) (2017): 149.

46 Roland Marchand, *Creating the Corporate Soul: The Rise of Public Relations and Corporate Imagery in American Big Business* (Berkeley: University of California Press, 1998), 202–203, 233–235, 238–239.

47 Trentmann, *Empire of Things*, 317–321; Douglas A. Galbi, "U.S. Annual Advertising Spending Since 1919," https://www.galbithink.org/ad-spending.htm.

48 Navon, "Truth in Advertising," 169–172.

49 Roland Marchand, *Advertising the American Dream: Making Way for Modernity, 1920–1940* (Berkeley: University of California Press, 1985), 117–163.

50 Navon, "Truth in Advertising," 158, 148.

51 Trentmann, *Empire of Things*, 317–321; Victoria de Grazia, "Changing Consumption Regimes in Europe, 1930–1970: Comparative Perspectives on the Distribution Problem," in *Getting and Spending*, ed. by Strasser, McGovern, and Judt, 66–68; Stearns, *Consumerism in World History*, 79–136.

52 Compare Warren Susman, *Culture as History: The Transformation of American Society in the Twentieth Century* (New York: Pantheon Books, 1985), 131–149.

53 Nathan O. Hatch, *The Democratization of American Christianity* (New Haven: Yale University Press, 1989), 128, 141–144; R. Laurence Moore, *Selling God: American Religion in the Marketplace of Culture* (New York: Oxford University Press, 1994), 17–20. See also Roger Finke and Rodney Stark, *The Churching of America, 1776–2005: Winners and Losers in Our Religious Economy* (New Brunswick, N.J.: Rutgers University Press, 2005).

54 Joel A. Tarr, *The Search for the Ultimate Sink: Urban Pollution in Historical Perspective* (Akron, Ohio: University of Akron Press, 1996), 323–333.

55 Martin V. Melosi, *Garbage in the Cities: Refuse, Reform, and the Environment: 1880–1980* (College Station: Texas A&M University Press, 1981), 169.

56 See, for example, prewar residential patterns in Gary, Indiana, a steel town founded by U.S. Steel in 1906. Andrew Hurley, *Environmental Inequalities: Class, Race, and Industrial Pollution in Gary, Indiana, 1945–1980* (Chapel Hill: University of North Carolina Press, 1995), 28–34.

57 Robert Bruegmann, *Sprawl: A Compact History* (Chicago: University of Chicago Press, 2005), 33–40; *Oxford English Dictionary*, "urban."

58 Simon Gunn and Susan C. Townsend, *Automobility and the City in Twentieth-Century Britain and Japan* (London: Bloomsbury Academic, 2019), 17–26; Thomas Zeller, *Driving Germany: The Landscape of the German Autobahn, 1930–1970*, trans. by Thomas Dunlap (New York: Berghahn Books, 2010), 48, 51–78.

59 Bruce L. Gardner, *American Agriculture in the Twentieth Century: How it Flourished and What it Cost* (Cambridge: Harvard University Press, 2002), 10–16.

60 Robert L. Mikkelsen and Thomas W. Bruulsema, "Fertilizer Use for Horticultural Crops in the U.S. during the 20th Century," *HortTechnology* 15(1) (2005): 24–30.

61 Edmund Russell, *War and Nature: Fighting Humans and Insects with Chemicals from World War I to Silent Spring* (Cambridge: Cambridge University Press, 2001), 74–88.

62 Kenneth Warren, *Bethlehem Steel: Builder and Arsenal of America* (Pittsburgh: University of Pittsburgh Press, 2008), 116–131.

63 Warren Dean, *Brazil and the Struggle for Rubber: A Study in Environmental History* (Cambridge: Cambridge University Press, 1987), 67–86; Greg Grandin, *Fordlandia: The Rise and Fall of Henry Ford's Forgotten Jungle City* (London: Icon, 2010); William Rosenau, *Corporations and Counterinsurgency* (Santa Monica, Cal.: RAND, 2009), 17–24; Gregg Mitman, "Forgotten Paths of Empire: Ecology, Disease, and Commerce in the Making of Liberia's Plantation Economy," *Environmental History* 22 (January 2017): 1–22.

64 Michael A. Bernstein, "The American Economy of the Interwar Era: Growth and Transformation from the Great War to the Great Depression," in *Calvin Coolidge and the Coolidge Era: Essays on the History of the 1920s*, ed. by John Earl Haynes (Washington, D.C.: Library of Congress, 1997), 192.

65 Brian Black, *Petrolia: The Landscape of America's First Oil Boom* (Baltimore: Johns Hopkins University Press, 2000), 60–81; Christopher W. Wells, *Car Country: An Environmental History* (Seattle: University of Washington Press, 2012), 178–181.

66 Bernstein, "The American Economy of the Interwar Era," 192.

67 Marc Reisner, *Cadillac Desert: The American West and Its Disappearing Water*, rev. and updated edn. (New York: Penguin Books, 1993), 1–168, 472–483, 486–487.

68 Melosi, *Garbage in the Cities*, 168–189; Martin V. Melosi, *The Sanitary City: Urban Infrastructure in America from Colonial Times to the Present* (Baltimore: Johns Hopkins University Press, 2000), 158–167.

69 Stéphane Frioux, "Settling Urban Waste Disposal Facilities in France c.1900–1940: A New Source of Inequality?," in *Environmental and Social Justice in the City: Historical Perspectives*, ed. by Geneviève Massard-Guilbaud and Richard Rodger (Cambridge: White Horse Press, 2011), 189–207. See also Sabine Barles, *L'invention des Déchets Urbains: France 1790–1970* (Seyssel, France: Champ Vallon, 2005).

70 Melosi, *Sanitary City*, 135–146.

71 Olegario, *Engine of Enterprise*, 139–140; Ali Kabiri, *The Great Crash of 1929: A Reconciliation of Theory and Evidence* (Basingstoke, Hampshire: Palgrave Macmillan, 2015), 188.

72 Christina D. Romer, "The Great Crash and the Onset of the Great Depression," *Quarterly Journal of Economics* 105(3) (1990): 597–624; Stephen G. Cecchetti and Georgios Karras, "Sources of Output Fluctuations During the Interwar Period: Further Evidence on the Causes of the Great Depression," *The Review of Economics and Statistics* 76(1) (1994): 80–102; "Steel Production Continues to Lag," *New York Times* (July 7, 1932): 31; Timothy F. Bresnahan and Daniel M. G. Raff, "Intra-Industry Heterogeneity and the Great Depression: The American Motor Vehicles Industry, 1929–1935," *Journal of Economic History* 51(2) (1991): 322; Trentmann, *Empire of Things*, 279.

7 *Stepping on the Gas*

1 See Darren Dochuk, *From Bible Belt to Sunbelt: Plain-Folk Religion, Grassroots Politics, and the Rise of Evangelical Conservatism* (New York: Norton, 2011).

2 Eric Schlosser, *Fast Food Nation: The Dark Side of the All-American Meal* (Boston: Mariner Books, 2001).

3 See Jan Van Bavel and David S. Reher, "The Baby Boom and Its Causes: What We Know and What We Need to Know," *Population and Development Review* 39(2) (2013): 257–88.

4 Robert J. Gordon, *The Rise and Fall of American Growth: The U.S. Standard of Living since the Civil War* (Princeton, N.J.: Princeton University Press, 2016), 18–19, 535–565.

5 See Gordon, *Rise and Fall of American Growth.*

6 See Kim Phillips-Fein, *Invisible Hands: The Businessmen's Crusade against the New Deal* (New York: Norton, 2010).

7 Frank Trentmann, *Empire of Things: How We Became a World of Consumers, from the Fifteenth Century to the Twenty-First* (New York: HarperCollins, 2016), 287–288.

8 See Darren E. Grem, *The Blessings of Business: How Corporations Shaped Conservative Christianity* (New York: Oxford University Press, 2016).

9 See Kevin Michael Kruse, *One Nation Under God: How Corporate America Invented Christian America* (New York: Basic Books, 2015); Jane Mayer, *Dark Money: The Hidden History of the Billionaires Behind the Rise of the Radical Right* (New York: Doubleday, 2016), 38–78; and Nancy MacLean, *Democracy in Chains: The Deep History of the Radical Right's Stealth Plan for America* (New York: Viking, 2017), 1–111.

10 Daniel Pope, *The Making of Modern Advertising* (New York: Basic Books, 1983),

254–257.

11 Paul Rutherford, *The New Icons? The Art of Television Advertising* (Toronto: University of Toronto Press, 1994), 38–44.

12 Kerwin C. Swint, *Dark Genius: The Influential Career of Legendary Political Operative and Fox News Founder Roger Ailes* (New York: Union Square Press, 2008), 1–30.

13 Daniel Yergin, *The Prize: The Epic Quest for Oil, Money, and Power* (New York: Simon & Schuster, 1991), 541–546; Christian Pfister, "The 'Syndrome of the 1950s' in Switzerland: Cheap Energy, Mass Consumption, and the Environment," in *Getting and Spending: European and American Consumer Societies in the Twentieth Century*, ed. by Susan Strasser, Charles McGovern, and Matthias Judt (New York: Cambridge University Press, 1998), 359–377.

14 Li Guoyu, *World Atlas of Oil and Gas Basins* (Chichester, West Sussex: Wiley-Blackwell, 2011), xvi., 158, 388.

15 Yergin, *The Prize*, 499–500.

16 Ibid., 425.

17 Ibid., 480, 488.

18 Ibid., 479–498.

19 Brian J. Cudahy, *Box Boats: How Container Ships Changed the World* (New York: Fordham University Press, 2006), 13–41.

20 Marc Levinson, *Outside the Box: How Globalization Changed from Moving Stuff to Spreading Ideas* (Princeton: Princeton University Press, 2020), 66.

21 Trentmann, *Empire of Things*, 326–337.

22 Ibid., 366–368, 372–399.

23 Christopher W. Wells, *Car Country: An Environmental History* (Seattle: University of Washington Press, 2012), 178–179.

24 John Sheail, "'Torrey Canyon': The Political Dimension," *Journal of Contemporary History* 42(3) (2007): 485–504; Robin J. Law, "The Torrey Canyon Oil Spill, 1967," in *Oil Spill Science and Technology: Prevention, Response, and Cleanup*, ed. by Mervin F. Fingas (Burlington, Mass.: Gulf Professional, 2011), 1103–1106.

25 Teresa Sabol Spezio, *Slick Policy: Environmental and Science Policy in the Aftermath of the Santa Barbara Oil Spill* (Pittsburgh, Pa.: University of Pittsburgh Press, 2018), xv–xvii, 1–4, 121–175.

26 Robert Bruegmann, *Sprawl: A Compact History* (Chicago: University of Chicago Press, 2005), 42–50.

27 Adam Rome, *The Bulldozer in the Countryside: Suburban Sprawl and the Rise of American Environmentalism* (Cambridge: Cambridge University Press, 2001).

28 Tom McCarthy, *Auto Mania: Cars, Consumers, and the Environment* (New Haven: Yale University Press, 2007), 116–121.

29 McCarthy, *Auto Mania*, 122–129.

30 Spencer R. Weart, *The Discovery of Global Warming*, rev. and exp. edn. (Cambridge: Harvard University Press, 2008), 1–37.

31 United States Census Bureau, *International Data Base*, https://www.census.gov/programs-surveys/international-programs/data/tools/international-data-base.html.

32 United States Department of Agriculture, Economic Research Service, drawing from the Food Expenditure Series by Eliana Zeballos and Wilson Sinclair, February 2022, https://www.ers.usda.gov/amber-waves/2020/november/average-share-of-income-spent-on-food-in-the-united-states-remained-relatively-steady-from-2000-to-2019/.

33 Paul Keith Conkin, *A Revolution Down on the Farm: The Transformation of American Agriculture Since 1929* (Lexington: University Press of Kentucky, 2008), 97–99; R. Douglas Hurt, *The Green Revolution in the Global South: Science, Politics, and Unintended Consequences* (Tuscaloosa: University of Alabama Press, 2020), 9–10.

34 Hurt, *Green Revolution in the Global South*, 8–101.

35 Ibid., 177–183; Conkin, *Revolution Down on the Farm*, 168–173.

36 Susan Strasser, *Waste and Want: A Social History of Trash* (New York: Metropolitan, 1999), 266–271.

37 Ibid., 271–274; Trentmann, *Empire of Things*, 624–627.

38 Reisner, *Cadillac Desert*, 120–305.

39 Philip Micklin, "The Future Aral Sea: Hope and Despair," *Environmental Earth Sciences* 75 (May 2016): 844; Iskandar Abdullaev, Kai Wegerich, and Jusipbek Kazbekov, "History of Water Management in the Aral Sea Basin," in *The Aral Sea Basin: Water for Sustainable Development in Central Asia*, ed. by Stefanos Xenarios, Dietrich Schmidt-Vogt, Manzoor Qadir, Barbara Janusz-Pawletta, and Iskandar Abdullaev (Abingdon, Oxon: Routledge, 2020).

40 Jennifer L. Derr, *The Lived Nile: Environment, Disease, and Material Colonial Economy in Egypt* (Stanford: Stanford University Press, 2019); Hesham Abd-El Monsef, Scot E. Smith, and Kamal Darwish, "Impacts of the Aswan High Dam After 50 Years," *Water Resources Management* 29(6) (2015): 1873–1885.

41 Mark Everard, *The Hydropolitics of Dams: Engineering or Ecosystems?* (London: Zed Books, 2013), 12–14, 20–22.

42 See Barbara L. Allen, *Uneasy Alchemy: Citizens and Experts in Louisiana's Chemical Corridor Disputes* (Cambridge: MIT Press, 2003).

43 Joel A. Tarr, *The Search for the Ultimate Sink: Urban Pollution in Historical*

Perspective (Akron, Ohio: University of Akron Press, 1996), 373–379; François Jarrige and Thomas Le Roux, *The Contamination of the Earth: A History of Pollutions in the Industrial Age*, trans. by Janice Egan and Michael Egan (Cambridge: MIT Press, 2020), 237–244; John Robert McNeill, *Something New Under the Sun: An Environmental History of the Twentieth-Century World* (New York: Norton, 2000), 122–135; "Film Developed in Polluted Water," *Washington Post*, November 17, 1971, A24; Nancy Langston, *Toxic Bodies: Hormone Disruptors and the Legacy of DES* (New Haven: Yale University Press, 2010), 1–16.

44 Jarrige and Le Roux, *Contamination of the Earth*, 238–244; Julien Boucher and Guillaume Billard, "The Challenges of Measuring Plastic Pollution," *Field Actions Science Reports*, Special Issue 19 (March 1, 2019), http://journals.openedition.org/factsreports/5319; Susanne Kühn, Elisa L. Bravo Rebolledo, and Jan A. van Franeker, "Deleterious Effects of Litter on Marine Life," in *Marine Anthropogenic Litter*, ed. by Melanie Bergmann, Lars Gutow, and Michael Klages (Cham, Switzerland: Springer, 2015), 75–116.

45 Sonja D. Schmid, *Producing Power: The Pre-Chernobyl History of the Soviet Nuclear Industry* (Cambridge: MIT Press, 2015), 17–40; McNeill and Engelke, *The Great Acceleration*, 27–32; James C. Williams, *Energy and the Making of Modern California* (Akron: University of Akron Press, 1997), 284–288 (quotation on 284).

8 Selling Everything

1 In 1911, a court antritrust decree divided Standard Oil into 35 companies, the largest of which, Standard Oil of New Jersey, alone was nearly as valuable as U.S. Steel.

2 David Halberstam, *The Fifties* (New York: Villard Books, 1993), 118.

3 Measured by brand value, according to the Kantar Group's "BrandZ™ Top 100 Most Valuable Global Brands 2020," 66, https://www.brandz.com/admin/uploads/files/2020_BrandZ_Global_Top_100_Report.pdf. The next nine companies in order were Apple, Microsoft, Google, Visa, Alibaba, Tencent, Facebook, McDonald's, and Mastercard. There are many measures of company value or size but Amazon always ranks at or near the top.

4 "Great Acceleration" of the twentieth century was probably first used in this sense by David Christian in *Maps of Time: An Introduction to Big History* (Berkeley: University of California Press, 2004), 440–464. Referring to the postwar era, J. R. McNeill, Peter Engelke, and others have popularized the term among environmental historians. See John Robert McNeill and Peter Engelke, *The Great Acceleration: An Environmental History of the Anthropocene Since 1945* (Cambridge: Harvard University Press, 2016), originally published as chapter

3 of Akira Iriye, ed., *Global Interdependence: The World After 1945* (Cambridge: Harvard University Press, 2014).

5 This and following paragraphs draw from Brad Stone, *The Everything Store: Jeff Bezos and the Age of Amazon* (New York: Little, Brown, 2013), 17–29.

6 James Wallace and Jim Erickson, *Hard Drive: Bill Gates and the Making of the Microsoft Empire* (New York: HarperBusiness, 1993), 6–7; Roger Lowenstein, *Buffett: The Making of an American Capitalist* (New York: Random House, 1995), 3.

7 Nick Hanauer, quoted in Alec MacGillis, *Fulfillment: Winning and Losing in One-Click America* (New York: Farrar, Straus, and Giroux, 2021), 193.

8 Stone, *Everything Store*, 139–159; Bezos married his first wife in a Catholic ceremony in 1993. J. K. Trotter, "What We Know, and Don't Know, About Jeff Bezos' Religious Beliefs," *Insider*, January 22, 2019, https://www.insider.com/what-we-know-about-jeff-bezos-religious-beliefs-after-divorce-2019-1. His mother Jaclyn Gise Jorgensen Bezos's upbringing inferred from her father's obituary: "Lawrence P. Gise," *Victoria Advocate* (Victoria, Texas), November 16, 1995.

9 Jessica Grogan, *Encountering America: Humanistic Psychology, Sixties Culture, and the Shaping of the Modern Self* (New York: Harper Perennial, 2013). On religion, see the following footnotes.

10 Randall J. Stephens, *The Devil's Music: How Christians Inspired, Condemned, and Embraced Rock 'n' Roll* (Cambridge: Harvard University Press, 2018).

11 Robert Wuthnow, *The Restructuring of American Religion: Society and Faith Since World War II* (Princeton: Princeton University Press, 1988), Wuthnow, *After Heaven: Spirituality in America Since the 1950s* (Berkeley: University of California Press, 1998); Steven P. Miller, *The Age of Evangelicalism: America's Born-Again Years* (New York: Oxford University Press, 2014); *The Transformation of the Christian Churches in Western Europe (1945–2000) / La Transformation des Églises Chrétiennes en Europe Occidentale*, ed. by Leo Kenis, Jaak Billiet, and Patrick Pasture (Leuven, Belgium: Leuven University Press, 2010).

12 See Daniel K. Williams, *God's Own Party: The Making of the Christian Right* (Oxford: Oxford University Press, 2010).

13 Frank Trentmann, *Empire of Things: How We Became a World of Consumers, from the Fifteenth Century to the Twenty-First* (New York: HarperCollins, 2016), 6.

14 See the brilliantly insightful Robert J. Gordon, *The Rise and Fall of American Growth: The U.S. Standard of Living since the Civil War* (Princeton, N.J.: Princeton University Press, 2016).

15 Kenneth Warren, *Big Steel: The First Century of the United States Steel Corporation, 1901–2001* (Pittsburgh, Pa.: University of Pittsburgh Press, 2001), 193–200, 214–215, 241, 245–258.

16 Jonathan Levy, *Ages of American Capitalism: A History of the United States* (New York: Random House, 2021), 544.

17 Warren, *Big Steel*, 309–339.

18 Janet Lowe, *Welch: An American Icon* (New York: Wiley, 2001); Peter Robison, *Flying Blind: The 737 Max Tragedy and the Fall of Boeing* (New York: Doubleday, 2021); Christopher Byron, *Testosterone Inc.: Tales of CEOs Gone Wild* (Hoboken, N.J.: Wiley, 2005).

19 The previous paragraphs draw upon Levy, *Ages of American Capitalism*, 544–632; John Ehrenreich, *Third Wave Capitalism: How Money, Power, and the Pursuit of Self-Interest Have Imperiled the American Dream* (Ithaca, N.Y.: Cornell University Press, 2016), 11; Youn Ki, "Large Industrial Firms and the Rise of Finance in Late Twentieth-Century America," *Enterprise and Society* 19(4) (2018): 903–945; Per H. Hansen, "From Finance Capitalism to Financialization: A Cultural and Narrative Perspective on 150 Years of Financial History," *Enterprise and Society* 15(4) (2014): 605–642.

20 Andrea Gabor, "Media Capture and the Corporate Education-Reform Philanthropies," in *Media Capture: How Money, Digital Platforms, and Governments Control the News*, ed. by Anya Schiffrin (New York: Columbia University Press, 2021), 117–140.

21 Gordon, *Rise and Fall of American Growth*, 605–634.

22 On American agriculture, see Gardner, *American Agriculture in the Twentieth Century*, 339–349.

23 Giovanni Federico, *Feeding the World: An Economic History of Agriculture, 1800–2000* (Princeton: Princeton University Press, 2008), 19–21, 28, 31–68.

24 Levy, *Ages of American Capitalism*, 637–640. See John Markoff, *What the Dormouse Said: How the Sixties Counterculture Shaped the Personal Computer Industry* (New York: Viking Penguin, 2005).

25 Levy, *Ages of American Capitalism*, 640–643.

26 Gordon, *Rise and Fall of American Growth*, 566–585.

27 For a short account of postwar global environmental change, see McNeill and Engelke, *The Great Acceleration*.

28 Nina Lakhani, Aliya Uteuova, and Alvin Chang, "Revealed: The True Extent of America's Food Monopolies, and Who Pays the Price," *Guardian* (London), July 14, 2021, https://www.theguardian.com/environment/ng-interactive/2021/jul/14/food-monopoly-meals-profits-data-investigation.

29 Michael Pollan, *The Omnivore's Dilemma: A Natural History of Four Meals* (New York: Penguin Press, 2006), 85–90.

30 Raoni Rajão, et al., "The Rotten Apples of Brazil's Agribusiness," *Science* 369(6501) (July 17, 2020): 246–248.

31 Paul Tullis, "How the World Got Hooked On Palm Oil," *Guardian* (London), February 19, 2019, https://www.theguardian.com/news/2019/feb/19/palm-oil-ingredient-biscuits-shampoo-environmental..

32 Dan Koeppel, *Banana: The Fate of the Fruit That Changed the World* (New York: Plume, 2009)

33 Callum Roberts, *The Unnatural History of the Sea* (Washington: Island Press, 2007), 168.

34 James E. Scarff, "The International Management of Whales, Dolphins, and Porpoises: An Interdisciplinary Assessment," *Ecology Law Quarterly* 6(2) (1977): 341–342, 344n86.

35 Kurkpatrick Dorsey, *Whales and Nations: Environmental Diplomacy on the High Seas* (Seattle: University of Washington Press, 2013).

36 Roberts, *Unnatural History*, 188–198, 203–213, 328–334, 338–339.

37 McNeill, *Something New*, 262–264; McNeill and Engelke, *The Great Acceleration*, 88–97; Elizabeth Kolbert, *The Sixth Extinction: An Unnatural History* (New York: Holt, 2014).

38 Pietra Rivoli, *The Travels of a T-Shirt in the Global Economy: An Economist Examines the Markets, Power, and Politics of World Trade*, 2nd edn. (Hoboken, New Jersey: Wiley, 2015).

39 Dana Thomas, *Fashionopolis: The Price of Fast Fashion—And the Future of Clothes* (New York: Penguin Press, 2019), 1–13.

40 See Susan Strasser, *Waste and Want: A Social History of Trash* (New York: Metropolitan, 1999), 161–201, 265–271; Kendra Smith-Howard, "Absorbing Waste, Displacing Labor: Family, Environment, and the Disposable Diaper in the 1970s," *Environmental History* 26(2) (April 2021): 207–230.

41 Ben Webster, "Amazon Destroys Lorry-loads of Unsold TVs and Computers," *The Times* (London), June 22, 2021.

42 Trentmann, *Empire of Things*, 622–675; Martin V. Melosi, *The Sanitary City: Urban Infrastructure in America from Colonial Times to the Present* (Baltimore: Johns Hopkins University Press, 2000), 240–258.

43 McNeill and Engelke, *The Great Acceleration*, 136–140; Anthony L. Andrady, *Plastics and Environmental Sustainability* (Somerset: Wiley, 2015), 145–254.

44 Nancy B. Grimm, et al., "Global Change and the Ecology of Cities," *Science* 319 (February 8, 2008): 756.

45 Lu Liu and Lina Meng, "Patterns of Urban Sprawl from a Global Perspective," *Journal of Urban Planning and Development* 146(2) (2020): 4020004; quotation on 2.

46 Grimm, et al., "Global Change and the Ecology of Cities," 756–760.

47 Shelby Gerking and Stephen F. Hamilton, "What Explains the Increased

Utilization of Powder River Basin Coal in Electric Power Generation?," *American Journal of Agricultural Economics* 90(4) (2008): 933–950; Shirley Stewart Burns, *Bringing down the Mountains: The Impact of Moutaintop Removal Surface Coal Mining on Southern West Virginia Communities* (Morgantown: West Virginia University Press, 2007), 9–18.

48 Joseph R. Gaudet, "The Energy Capital of the World: A History of Grass, Oil, and Coal in the Powder River Basin" (Ph.D. diss., University of Michigan, 2019), 220–281.

49 Shrabani Mukherjee, "Environmental Costs of Coal-Based Thermal Power Generation in India: Notion and Estimation," in *Environmental Scenario in India: Successes and Predicaments*, ed. by Sacchidananda Mukherjee and Debashis Chakraborty (London: Routledge, 2014).

50 Christopher J. Rhodes, "The Global Oil Supply—Prevailing Situation and Prognosis," *Science Progress* 100(2) (2017): 231–240; Johannes Peter Gerling, and Friedrich-Wilhelm Wellmer, "Wie Lange Gibt Es Noch Erdöl und Erdgas? Reserven, Ressourcen und Reichweiten," *Chemie in Unserer Zeit* 39(4) (2005): 236–245.

51 International Tanker Owners Pollution Federation, *Oil Tanker Spill Statistics 2020* (London: ITOPF, 2021), 6.

52 "Crude Calamities—the Biggest Offshore Oil Spill Disasters," *Progressive Digital Media Oil and Gas, Mining, Power, CleanTech and Renewable Energy News*, September 11, 2014; Graeme MacDonald, "Containing Oil: The Pipeline in Petroculture," in *Petrocultures: Oil, Politics, Culture*, ed. by Sheena Wilson, Adam Carlson, and Imre Szeman (Montreal: McGill–Queen's University Press, 2017), 62, 76n6.

53 Katherine Kornei, "Here Are Some of the World's Worst Cities for Air Quality," *Science* (March 21, 2017).

54 McNeill, *Something New Under the Sun*, 99–102.

55 See Stephen O. Andersen and K. Madhava Sarma, *Protecting the Ozone Layer: The United Nations History* (London: Earthscan, 2002), 1–41.

56 Scott C. Doney, "The Dangers of Ocean Acidification," *Scientific American* 294(3) (March 2006): 58–65.

57 See Weart, *The Discovery of Global Warming*, 63–113, 177–196.

9 The Rise and Globalization of Environmentalism

1 Michael B. Smith, "'Silence, Miss Carson!' Science, Gender, and the Reception of 'Silent Spring,'" *Feminist Studies* 27(3) (2001): 733–752; Maril Hazlett, "'Woman vs. Man vs. Bugs': Gender and Popular Ecology in Early Reactions to Silent Spring," *Environmental History* 9(4) (2004): 701–729.

2 Linda J. Lear, *Rachel Carson: Witness for Nature* (New York: Holt, 1997), 7–80.

3 Mark Stoll, *Inherit the Holy Mountain: Religion and the Rise of American Environmentalism* (Oxford: Oxford University Press, 2015), 172–201.

4 Yaakov Garb, "Rachel Carson's *Silent Spring*," *Dissent* 42 (Fall 1995): 539–546.

5 Jean Gartlan, *Barbara Ward: Her Life and Letters* (London: Continuum, 2010), 1–25.

6 Ibid., 27–95.

7 Ibid., 97–204. See also Barbara Ward, "Only One Earth, Stockholm 1972," and Brian Johnson, "The Duty to Hope: A Tribute to Barbara Ward," both in *Evidence for Hope: The Search for Sustainable Development*, ed. by Nigel Cross (London: Taylor & Francis Group, 2003), 3–9, 10–18; and Cedric Pugh, "Introduction," in *Sustainability, the Environment and Urbanization*, ed. by Cedric D. J. Pugh (London: Earthscan, 1996), 15–16.

8 Fox, *American Conservation Movement*, 148–182, 190–199.

9 Paul S. Sutter, *Driven Wild: How the Fight against Automobiles Launched the Modern Wilderness Movement* (Seattle: University of Washington Press, 2002).

10 Neil M. Maher, *Nature's New Deal: The Civilian Conservation Corps and the Roots of the American Environmental Movement* (Oxford: Oxford University Press, 2008); Douglas Brinkley, *Rightful Heritage: Franklin D. Roosevelt and the Land of America* (New York: Harper, 2016); T. H. Watkins, *Righteous Pilgrim: The Life and Times of Harold L. Ickes, 1874–1952* (New York: H. Holt, 1990).

11 Donald Worster, *Dust Bowl: The Southern Plains in the 1930s* (New York: Oxford University Press, 1979); Paul B. Sears, *Deserts on the March* (Norman: University of Oklahoma Press, 1935); Paul Warde, Libby Robin, and Sverker Sörlin, *The Environment: A History of the Idea* (Baltimore: Johns Hopkins University Press, 2018), 35, 73–78.

12 Ibid., 25–72.

13 Howard Brick, *Age of Contradiction: American Thought and Culture in the 1960s* (New York: Twayne, 1998), 124–131.

14 Warde, Robin, and Sörlin, *Environment*, 6–24; Rachel Carson, *Silent Spring* (Boston: Houghton Mifflin, 1962), 6.

15 See Bruce Clarke, *Gaian Systems: Lynn Margulis, Neocybernetics, and the End of the Anthropocene* (Minneapolis: University of Minnesota Press, 2020).

16 Brick, *Age of Contradictions*, 131–136.

17 Warde, Robin, and Sörlin, *Environment*, 8–14; Osborn quoted on 14.

18 Ralph H. Lutts, "Chemical Fallout: Rachel Carson's *Silent Spring*, Radioactive Fallout, and the Environmental Movement," *Environmental Review* 9 (Fall 1985): 211–225; Jimmie M. Killingsworth and Jacqueline S. Palmer, "Millennial Ecology: The Apocalyptic Narrative from *Silent Spring* to *Global Warming*," in *Green Culture: Environmental Rhetoric in Contemporary America*, ed. by Carl G. Herndl and Stuart C. Brown (Madison: University of Wisconsin Press, 1996). See also Jacob Darwin Hamblin, *Arming Mother Nature: The Birth of Catastrophic Environmentalism* (Oxford University Press, 2013).

19 Stewart L. Udall, *The Quiet Crisis* (New York: Holt, Rinehart and Winston, 1963).

20 Mark W. T. Harvey, *A Symbol of Wilderness: Echo Park and the American Conservation Movement* (Albuquerque: University of New Mexico Press, 1994); Byron E. Pearson, *Still the Wild River Runs: Congress, the Sierra Club, and the Fight to Save Grand Canyon* (Tucson: University of Arizona Press, 2002); Marc Reisner, *Cadillac Desert: The American West and Its Disappearing Water*, rev. and updated edn. (New York: Penguin Books, 1993), 241–254; Joachim Radkau, *The Age of Ecology: A Global History*, trans. by Patrick Camiller (Cambridge: Polity, 2014), 137–148.

21 Thomas Raymond Wellock, *Critical Masses: Opposition to Nuclear Power in California, 1958–1978* (Madison: University of Wisconsin Press, 1998); Henry F. Bedford, *Seabrook Station: Citizen Politics and Nuclear Power* (Amherst: University of Massachusetts Press, 1990); Joan Aron, *Licensed to Kill? The Nuclear Regulatory Commission and the Shoreham Power Plant* (Pittsburgh: University of Pittsburgh Press, 1997); J. Samuel Walker, *Three Mile Island: A Nuclear Crisis in Historical Perspective* (Berkeley: University of California Press, 2004).

22 Fox, *American Conservation Movement*, 250–329.

23 Adam Rome, *The Bulldozer in the Countryside: Suburban Sprawl and the Rise of American Environmentalism* (Cambridge: Cambridge University Press, 2001), 119–128; Brick, *Age of Contradictions*, 1–22, 113–119; Richard H. Pells, *The Liberal Mind in a Conservative Age: American Intellectuals in the 1940s and 1950s* (New York: Harper & Row, 1985), 183–261; Roderick Nash, *Wilderness and the American Mind*, 5th edn. (New Haven: Yale University Press, 2014), 251–254; Catherine L. Albanese, *Nature Religion in America: From the Algonkian Indians to the New Age* (Chicago: University of Chicago Press, 1980), 153–198; Russell Duncan, "The Summer of Love and Protest: Transatlantic Counterculture in the 1960s," in *The Transatlantic Sixties: Europe and the United States in the Counterculture Decade*, ed. by Grzegorz Kosc, Clara Juncker, Sharon Monteith, and Britta Waldschmidt-Nelson (Bielefeld: Transcript Verlag, 2013), 144–173.

24 Andrew Jamison, Ron Eyerman, and Jacqueline Cramer, *The Making of the*

New Environmental Consciousness: A Comparative Study of the Environmental Movements in Sweden, Denmark and the Netherlands (Edinburgh: Edinburgh University Press, 1990), 18–22; Martin Kylhammar, *Nils Dahlbeck: En Berättelse om Svensk Natur och Naturvårdshistoria* (Stockholm: Carlsson, 1992), 56–60; Frank Graham, Jr., *Since Silent Spring* (Boston: Houghton Mifflin, 1970), 119–121. See also Lennart J. Lundqvist, *Environmental Policies in Canada, Sweden, and the United States: A Comparative Overview* (Beverly Hills, Calif.: Sage, 1974).

25 Graham, *Since Silent Spring*, 81–86, 117–119; John Scheail, *An Environmental History of Twentieth-Century Britain* (Basingstroke, Hampshire: Palgrave, 2002), 222, 235–45; G. R. Conway, D. G. R. Gilbert, and J. N. Pretty, "Pesticides in the UK: Science, Policy and the Public," in *Britain Since 'Silent Spring': An Update on the Ecological Effects of Agricultural Pesticides in the UK*, ed. by D. J. L. Harding (London: Institute of Biology, 1988).

26 Jamison, et al., *Making of the New Environmental Consciousness*, 129, 133, 136; Graham, *Since Silent Spring*, 21, 23, 51–52, 81, 88–89; C. J. Briejèr, *Zilveren Sluiers en Verborgen Gevaren: Chemische Preparaten die het Leven Bedreigen* (Leiden: Sijthoff, 1968), ch. 12; Jacqueline Cramer, *De Groene Golf: Geschiedenis en Toekomst van de Milieubeweging* (Utrecht: Jan van Arkel, 1989), 19–23, 36; J. L. van Zanden and S. W. Verstegen, *Groene Geschiedenis van Nederland* (Utrecht: Het Spectrum, 1993), 77–78.

27 Winfried Kösters, *Umweltpolitik: Themen, Funktionen, Zuständigkeiten* (Munich: Olzog, 1997), 12, 33; Kai F. Hünemörder, *Die Frühgeschichte der globalen Umweltkrise und die Formierung der deutschen Umweltpolitik (1950–1973)* (Stuttgart: Steiner, 2004), 121–126; Andrea Westermann, "Chemisierung der Landwirtschaft und Problemwahrnehmung durch den hauptamtlichen Naturschutz in den 1960er und 1970er Jahren: Die Rezeption von Carsons 'Der stumme Frühling' (1962) als ein exemplarischer Fall für das Verhältnis von Wissenschaft und Öffentlichkeit" (unpublished paper); Raymond H. Dominick, *The Environmental Movement in Germany: Prophets and Pioneers, 1871–1971* (Bloomington: Indiana University Press, 1992), 122–123, 152, 156.

28 Jamison et al., *Making of the New Environmental Consciousness*, 73.

29 Radkau, *Age of Ecology*, 79–113, 365–377; Frank Uekotter, *The Greenest Nation? A New History of German Environmentalism* (Cambridge: MIT Press, 2014), 59–100.

30 Joel Bartholemew Hagen, *An Entangled Bank: The Origins of Ecosystem Ecology* (New Brunswick, N.J.: Rutgers University Press, 1992), 100–121.

31 Warde, Robin, and Sörlin, *Environment*, 21–22, 47–72, 108–119. See, for example, Graham M. Turner, "A Comparison of *The Limits to Growth* with 30 Years

of Reality," *Global Environmental Change* 18(3) (August 2008): 397–411; Gaya Herrington, "Update to *Limits to Growth*: Comparing the World3 Model with Empirical Data," *Journal of Industrial Ecology* 25(3) (2021): 614–626.

32 Armin Rosencranz, Shyam Divan, and Antony Scott, "Legal and Political Repercussions in India," and Josée van Eijndhoven, "Disaster Prevention in Europe," in *Learning from Disaster: Risk Management After Bhopal*, ed. by Sheila Jasanoff (Philadelphia: University of Pennsylvania Press, 1994), 44–65, 113–132.

33 John McCormick, *Reclaiming Paradise: The Global Environmental Movement* (Bloomington: Indiana University Press, 1989), 137–143; Joachim Radkau, *Nature and Power: A Global History of the Environment*, trans. by Thomas Dunlap (Cambridge: Cambridge University Press, 2008), 264–265; Radkau, *Age of Ecology*, 148–155, 158–162, 167–169; Uekotter, *Greenest Nation*, 113–155.

34 Mark Stoll, "Les influences religieuses sur le mouvement écologiste français," in *Une protection de la nature et de l'environnement à la française*, ed. by Charles-François Mathis and Jean-François Mouhot (Seyssel, France: Champ Vallon, 2013).

35 Dominick, *Environmental Movement in Germany*, 207–208. See also H. Merbitz, "Wenn der Acker zur Fabrik wird: Eine Diskussion über unbekannte Gefahren des chemischen Pflanzenschutzes," *Christ und Welt* 16(19) (May 10, 1963):30; Richard Kaufmann, "Sind sie wirklich harmlos? Die Frage der Pflanzenschutzmittel im Licht eines amerikanischen Regierungsberichts," *Christ und Welt* 16(23) (June 7, 1963): 21; and Hans-J. Wasserburger, "Das Gift im Boden: Schädlingsbekämpfung muß 'integriert' erfolgen," *Christ und Welt* 16(46) (November 15, 1963): 22. The controversy spilled over into side columns and the letters to the editor.

36 Julia E. Ault, "Defending God's Creation? The Environment in State, Church and Society in the German Democratic Republic, 1975–1989," *German History* 37(2) (June 2019): 205–226.

37 Radkau, *Age of Ecology*, 80, 85.

38 Jamison, et al., *Making of the New Environmental Consciousness*, 13–53, 66–114; Peder Anker, *The Power of the Periphery: How Norway Became an Environmental Pioneer for the World* (Cambridge: Cambridge University Press, 2020).

39 Elizabeth D. Blum, *Love Canal Revisited: Race, Class, and Gender in Environmental Activism* (Lawrence: University Press of Kansas, 2008).

40 Barbara L. Allen, *Uneasy Alchemy: Citizens and Experts in Louisiana's Chemical Corridor Disputes* (Cambridge: MIT Press, 2003).

41 See Robert D. Bullard, *Dumping in Dixie: Race, Class, and Environmental Quality* (Boulder: Westview Press, 1990); Ellen Griffith Spears, *Baptized in PCBs: Race,*

Pollution, and Justice in an All-American Town (Chapel Hill: University of North Carolina Press, 2014); Linda Lorraine Nash, *Inescapable Ecologies: A History of Environment, Disease, and Knowledge* (Berkeley: University of California Press, 2006).

42 Radkau, *Age of Ecology*, 377–401.

43 Ibid., 377–382; Pope Francis, *Laudato Si': On Care for Our Common Home* (Huntington, Ind.: Our Sunday Visitor, 2015).

44 Barbara Ward and René J. Dubos, *Only One Earth: The Care and Maintenance of a Small Planet* (New York: Norton, 1972), 191–208.

45 Ibid., 158–159.

46 Jennifer Krill, "Rainforest Action Network," in *Good Cop/Bad Cop: Environmental NGOs and Their Strategies Toward Business*, ed. by Thomas P. Lyon (Washington, D.C.: RFF, 2010), 208–220.

47 Pamela S. Chasek and Lynn M. Wagner, "An Insider's Guide to Multilateral Environmental Negotiations since the Earth Summit," in *The Roads from Rio: Lessons Learned from Twenty Years of Multilateral Environmental Negotiations*, ed. by Pamela Chasek and Lynn M. Wagner (New York: Routledge, 2012), 1–15; Radkau, *Age of Ecology*, 386–390.

48 Bill McKibben, *The End of Nature* (New York: Random House, 1989).

49 Spencer R. Weart, *The Discovery of Global Warming*, rev. and expanded edn. (Cambridge: Harvard University Press, 2008), 138–176.

50 Keith E. Peterman, "Contentious Journey from Rio to Paris and the Path Beyond," in *Global Consensus on Climate Change: Paris Agreement and the Path Beyond*, ed. by Keith E. Peterman, Gregory P. Foy, and Matthew R. Cordes (Washington, D.C.: American Chemical Society, 2019), 1–9.

51 Ward and Dubos, *Only One Earth*, 156–170.

52 Isabelle Garzon, *Reforming the Common Agricultural Policy: History of a Paradigm Change* (Basingstoke: Palgrave Macmillan, 2006), 21–40; Venus Bivar, *Organic Resistance: The Struggle Over Industrial Farming in Postwar France* (Chapel Hill: University of North Carolina Press, 2018).

53 Matthew Reed, *Rebels for the Soil: The Rise of the Global Organic Food and Farming Movement* (London: Earthscan, 2010); William Lockeretz, ed., *Organic Farming: An International History* (Wallingford: CABI, 2007).

54 Darren Dochuk, *Anointed with Oil: How Christianity and Crude Made Modern America* (New York: Basic Books, 2019); Radkau, *Age of Ecology*, 270–271.

55 Hartmut Berghoff, "Shades of Green: A Business-History Perspective on Eco-Capitalism"; and Hugh S. Gorman, "The Role of Businesses in Constructing Systems of Environmental Governance," both in *Green Capitalism? Business and the Environment in the Twentieth Century*, ed. by Hartmut Berghoff and Adam

Rome (Philadelphia: University of Pennsylvania Press, 2017); Radkau, *Age of Ecology*, 270–271.

56 See Jane Mayer, *Dark Money: The Hidden History of the Billionaires Behind the Rise of the Radical Right* (New York: Doubleday, 2016).

57 A play on words on the famous message of American Commodore Perry after the Battle of Lake Erie in the War of 1812: "We have met the enemy and he is ours."

58 Michael E. Mann, *The New Climate War: The Fight to Take Back the Planet* (New York: PublicAffairs, 2021).

59 The essential book is Naomi Oreskes and Erik M. Conway, *Merchants of Doubt: How a Handful of Scientists Obscured the Truth on Issues from Tobacco Smoke to Global Warming* (New York: Bloomsbury Press, 2010).

60 Gorman, "Role of Businesses," 44–45.

61 Berghoff, "Shades of Green," 24.

Conclusion: Profit — Capitalism and Environment

1 Inger L. Stole, "Philanthropy as Public Relations: A Critical Perspective on Cause Marketing," *International Journal of Communication* 2 (2008): 20–40; Al Ries and Laura Ries, *The Fall of Advertising and the Rise of PR* (New York: HarperBusiness, 2004), xi–xxi.

2 Homi Kharas and Wolfgang Fengler, "Global Poverty is Declining but Not Fast Enough," *Future Development* [BLOG], November 17, 2017, https://www.brookings.edu/blog/future-development/2017/11/07/global-poverty-is-declining-but-not-fast-enough/.

3 Jonathan Watts, "Blue-Sky Thinking: How Cities Can Keep Air Clean after Coronavirus," *Guardian* (London), June 7, 2020.

4 "IKEA Executive on Why the West Has Hit 'Peak Stuff,'" *All Things Considered,* NPR Radio, January 22, 2016.

Index